Logging Towboats and Boom Jumpers

The Story of O.A. Harkness

Logging "Admiral" and Mechanical "Wizzard"

Roger Allen Moody

Published by North Country Press
Unity, Maine

ISBN 978-1-943424-41-2

Library of Congress Number 2018957360

Photo credits: shown by each photo

On the cover: **Great Northern Paper Company photo "boom jumper" GNP Co. *No. 41* at North Twin Lake, 1949**

Great Northern Paper Company Records, SpC MS 0210 Box 24, Folder 27, Raymond H. Fogler Library Special Collections Department, University of Maine, Orono, Maine

Printed in the United States of America

North Country Press
Unity, Maine

Contents

Acknowledgments

The author would like to acknowledge the important and valuable contributions of Vinton Orris "V.O." Harkness, grandson of Orris Albert "O.A." Harkness, for providing direct access to family records and ship models as resource materials for this book. Those family records provided information not available from any other source. Additionally, V.O.'s personal support and enthusiasm for this book added to the sheer enjoyment of writing about the work, creativity, and productivity of his fascinating grandfather.

Gratitude goes to Herbert Crosby of the Maine Forest and Logging Museum, and to Terence Harper for sharing their research of and enthusiasm for Maine's logging industry history. Herb Crosby is a University of Maine Professor Emeritus of Mechanical Engineering Technology, having taught machine design, thermodynamics, and student capstone design projects. He helped lead the restoration of the circa 1910 Lombard steam log hauler at the Maine Forest and Logging Museum to working condition. Terry Harper teaches drafting technology courses at the Presque Isle Regional Career and Technical Center. Working with the Allagash Alliance, he has spent more than 20 years researching and reclaiming the two abandoned locomotives at the historic Eagle Lake and West Branch Railroad site deep in some of Maine's wildest terrain.

Special appreciation goes to my spouse, Audrey Moody, both for her encouragement and support during the process of researching and writing this book and for her thorough manuscript edits.

Thanks also to reader Craig White of the Owl & Turtle Book Store in Camden, Maine, for his insights and suggestions which materially improved this work.

Camden, Maine
June 1, 2018

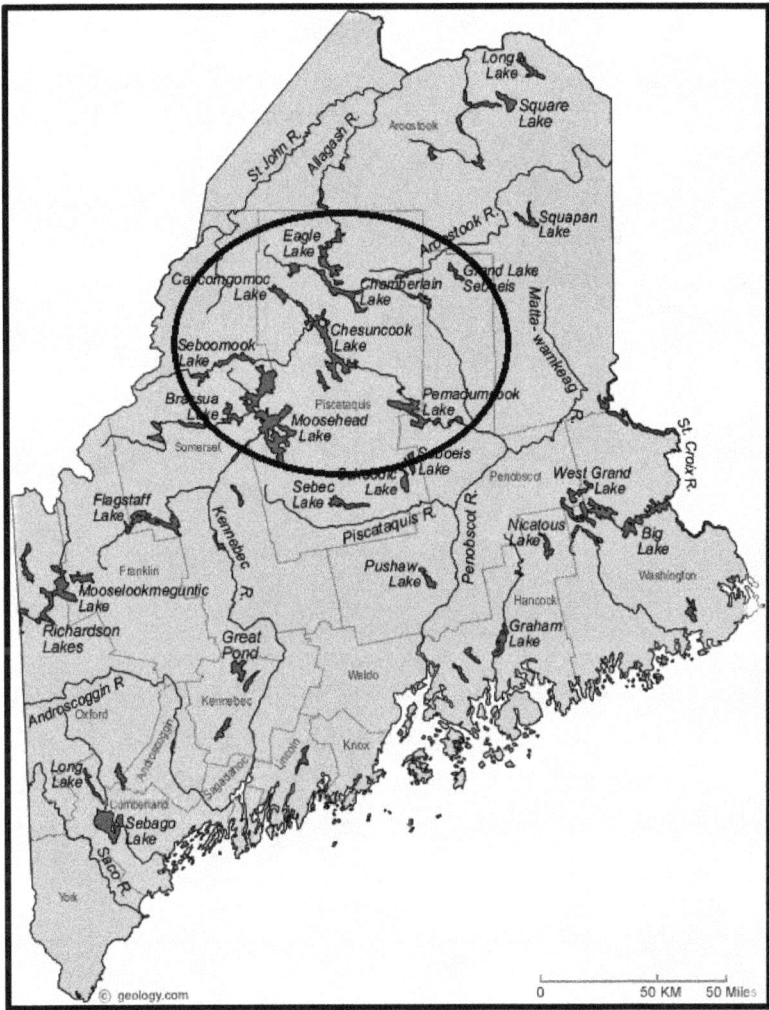

1932 Great Northern Paper Company Map

Great Northern Paper Company Records, SpC MS 0210, Raymond H. Fogler Library
Special Collections Department, University of Maine, Orono, Maine

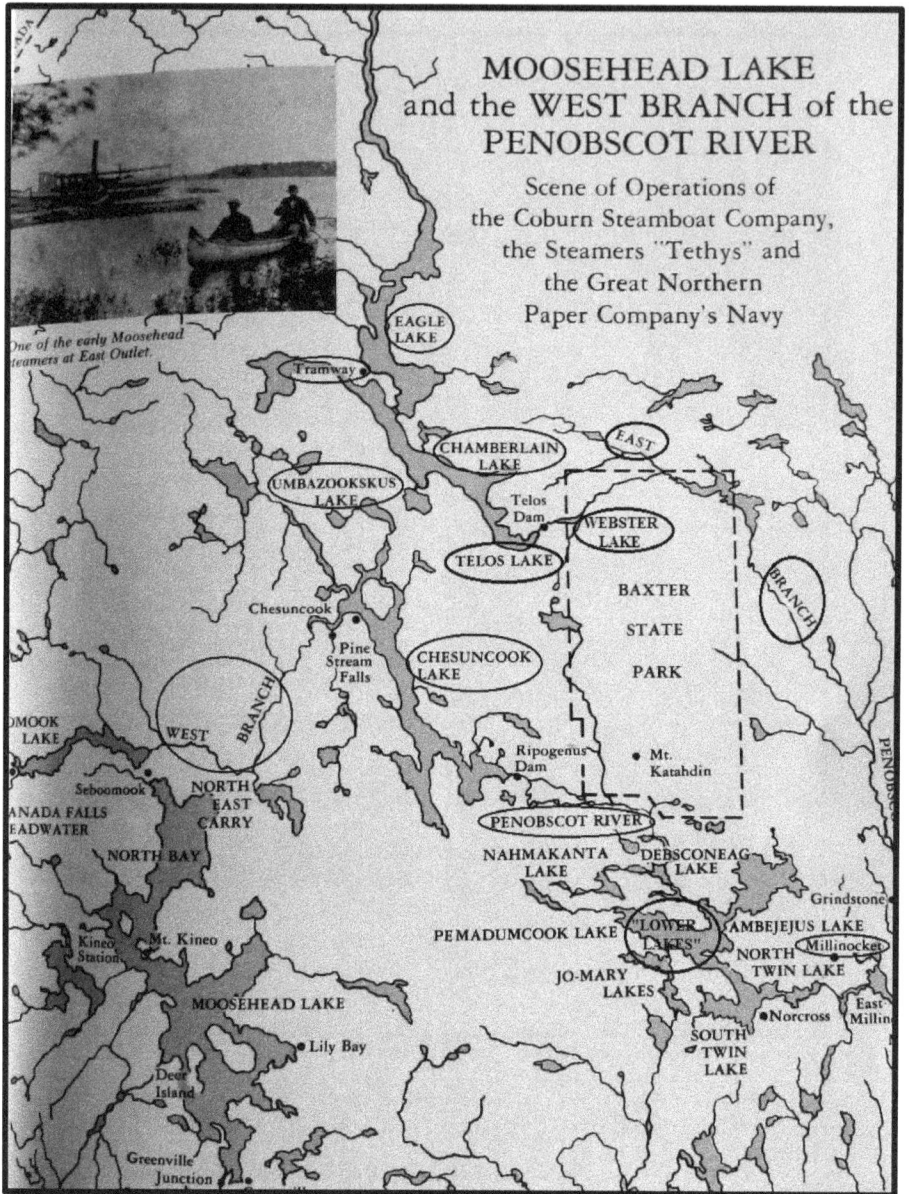

MOOSEHEAD LAKE
and the WEST BRANCH of the
PENOBSCOT RIVER

Scene of Operations of
the Coburn Steamboat Company,
the Steamers "Tethys" and
the Great Northern
Paper Company's Navy

One of the early Moosehead
steamers at East Outlet.

Adapted from Great Northern Paper Company Records, SpC MS 0210, Raymond H.
Fogler Library Special Collections Department, University of Maine, Orono, Maine

Introduction

Much has been written over the years about the Great Northern Paper Company (GNP Co.), located on the West Branch of Maine's Penobscot River near its confluence with the East Branch and the main stem of the Penobscot River itself. Its loggers, logging operations, mills, use of industrial technology, evolving systems for moving logs via the West and East Branches from forest to mills, and privately-owned dams (which initially controlled water flow for springtime log drives and later were developed into the largest privately-owned hydroelectric system in the United States) have been the subject of much research and many books.

The West Branch drains spruce and fir forests of the southerly part of Maine's north woods and drops 1,174 feet (358 m) from its headwaters along the Maine-Quebec border to Medway at the confluence of the West and East Branches of the Penobscot River. Logging of the West Branch region began in 1828 when logs were loaded onto sleds and towed to the river by oxen and horses during winter. In the spring when the river swelled with the melting of winter's snow and ice, log drives would float the logs downstream as far as the tidal waters of Bangor, Maine, for sorting, sawing, and shipment to ports throughout the eastern United States. Harvesting tree-length logs on the West Branch for the sawmill industry peaked in 1872, and in the late 1890s the focus of timber harvesting shifted to supplying 4' pulpwood for the paper-making industry.

The Maine legislature chartered the Northern Development Company in 1897, granting it the authority to generate and market hydroelectricity for manufacturing. The company was renamed Great Northern Paper Company in 1899. Its Millinocket mill, then the largest paper mill in the world, began operation in 1900 and within a year had captured 12 percent of the United States newsprint market. [A new, large GNP Co. mill was also built seven miles downstream at East Millinocket in 1906.] GNP Co. held that market share for 16 years by increasing its competitive efficiency, especially with the electricity output through the series of hydroelectric dams it built along the West Branch. In 1903 the Maine legislature gave exclusive control of the West Branch above Shad Pond (near Millinocket) to GNP Co.'s subsidiary, the West Branch Driving and Reservoir Dam Company. [1]

From the company's initial forestlands of 250,000 acres, its land holdings increased to 2,100,000 acres. From these and other lands, most of the timber was sent down the waters of East and West Branches to become pulp fiber for papermaking. As logging operations moved ever farther from the two main stems of the river, mechanical log haulers, a tramway, and a short-line railroad became a part of the technology used for bringing pulp logs from adjacent watersheds into those of the East and West Branches of the Penobscot. [2]

Great Northern existed as a paper-making entity until about 2003.

As Superintendent of Mechanical Equipment for GNP Co., Orris Albert "O.A." Harkness played a key role in developing and utilizing evolving mechanical and industrial technologies to move logs ever more efficiently from woodlands to the mills. He designed, built, and maintained a fleet of tow boats and "boom jumpers," and kept continuous-tracked log haulers, sophisticated conveyors, a 3000' tramway, crawler tractors, a specialized train, motor vehicles, and generators in full operation. His talents combined both the practical ingenuity and long-term vision necessary to keep the motorized equipment performing their tasks with increasing effectiveness and reliability.

Written for those who are intrigued by history, this book is the story of how O.A. Harkness fulfilled a unique and important role which captured his mechanical genius as the GNP Co.'s "Wizzard" [sic] of motor power. It especially focuses on his role as GNP Co.'s boat fleet "Admiral" who with great pride designed, built, and maintained company watercraft for log driving operations. Additionally, it contains expansive information about the many aspects of logging in the first half of the 20th century.

Chapter 1

O.A. Harkness – His Life in Summary

O.A. Harkness was called, in *The Northern* magazine published by the Social Services Division of his employer the Great Northern Paper Company (GNP Co.), a "mechanical genius." In 1927, the publication described him this way: "of all the men in the Spruce Wood Department, none travels farther or is in greater demand than O.A. Harkness. He may be seen anywhere [*sic*], but not for long, as he is constantly covering the whole territory where gasoline, oil or coal is used for motive power. It is difficult to hold him long enough for an interview about his greatest interest – his inland fleet of boats. This fleet he has built up for handling pulpwood on the rivers and lakes of northern Maine. He takes pride in his boats, and well he may, for he is responsible for building as well as operating them. Last year the building of the *West Branch No. 2* crowned the construction work with its greatest prize. – Mr. Harkness was dubbed 'Admiral' and the *West Branch No. 2* became his flag ship." (A detailed description and photo of the vessel is contained in Chapter 3.)

The Northern typically contained a column of brief personal observations and society news called "Here and There," and the July 1923 edition contains a brief item with the headline "THE WIZZARD [*sic*] OF MOTOR POWER." It reported "A few months ago we had occasion to call attention to the fact that one of our number was the first to travel [to] Caucomgomoc Lake by auto. He has more recently performed a couplet of fetes [*sic*] which greatly add to his laurels. Within a few weeks, he rode from Kineo to Greenville and only a few days from that time rode by boat into Island Falls through the main street - this man is Harkness."

O.A. Harkness

The Northern, May 1925 p.8 Great Northern Paper Company Records, SpC MS 0210, Box 21, Raymond H. Fogler Library Special Collections Department, University of Maine, Orono, Maine

It is very appropriate that Mr. O.A. Harkness as Superintendent of Motor Equipment should be the greatest traveler we have in the Northern territory. He is always headed for somewhere, and doesn't stay long when he arrives. Company men spending the night at a hotel in Houlton have found a note at the desk left by Mr. Harkness as he sped through the town at 3:00 a.m. It is no wonder that his Franklin registers more than 25,000 miles or more during a season. He is in great demand and called in every direction. The light plant at 40 mile, the loader at Washburn, a tractor at Grindstone, a motor boat at Norcross and a steam boat at Chesuncook Lake may all need his attention at one time.

He has been with the Great Northern Paper Company for about ten years. During that time, he has had charge of all motor equipment, including the repair shop at Greenville, the fleet of trucks, tractors, touring cars, steamboats, motor boats, and stationary engines such as light-plants, pumping stations, and conveyors. In addition, he usually has a boat or two under construction and affair [*sic*] number of small buildings going up or undergoing repair.

For this work, Mr. Harkness is well qualified by his natural aptitude and past experience. His first job as a boy was helping on a little steamboat that took parties out on Megunticook Lake, Camden, Maine.

A few years later he owned a forty-foot steamboat with which he did this work. The time spent in the experimental department of the Walworth Manufacturing Company of South Boston, the experience he gained at Bath Iron Works, together with the time spent in building boats and cottages and in the construction of engines and boilers, all prepared him in a unique way for the important positions he has since held. Before coming to the Great Northern Paper Company, Mr. Harkness held a position with the Eastern Manufacturing Company for about fifteen years. This work kept him busy for some time at Eagle and Chamberlain Lakes, building boats and looking after the Eagle Lake Tramway and the log-haulers.

Mr. and Mrs. Harkness have two children, both of whom are graduates of the University of Maine. Their son Vinton Orris Harkness was graduated in the class of 1922 and is now an employee of Fairbanks, Morse & Company of Boston.

The accompanying picture is a good likeness of Mr. Harkness as he is now [see above]. Friends of former days will not recognize him without the heavy dark moustache and beard [see below] that used to distinguish him. However, he is the same calm man of affairs, sure of a cordial welcome wherever he goes. His even disposition and his consideration for others have ever won for him the friendship of his associates and the respect of the men who work for him." [3]

A younger O.A. Harness with his heavy dark mustache and beard
Vinton Orris "V.O." Harkness, Jr. (O.A. Harkness's grandson) Collection

O.A. Harkness – Family History

The Harkness ancestors were among the earliest European settlers at Goose River (now the town of Rockport, Maine, and specifically Rockport Harbor) which was, until 1891, part of the town of Camden. John Harkness (1750-1806) was a significant landowner on the west side of the harbor who married Elizabeth Ott. He was elected as Camden's initial first selectman after the incorporation of the town in 1791. Family members remained in Rockport through the several generations preceding the birth of O.A. Harkness.

Orris Albert Harkness was born on July 6, 1869, in Camden, Maine, to Ephraim T. Harkness (1845-1930) and Eliza Ann Woster (1849-1930). Ephraim and Eliza lived in Rockport near the corner of present-day Pascal Avenue and Commercial Street, as evidenced by an 1861 deed exchanging land with Camden School District No. 18 to allow for the construction of a school house (the Hoboken School) at that location.

O.A. Harkness married Anna Eliza Fernald in 1891, and the 1900 U.S. Census shows the couple living in Lincolnville at the Megunticook

Lake farm of Lucy Fernald, Anna Eliza's mother. Anna Eliza was born in Lincolnville, Maine, on December 14, 1862, and died in Lincolnville in 1928 [buried in the Lower Cemetery]. Their children were Elizabeth A. Harkness (born 1900) and Vinton Orris Harkness (1898-1979). [Vinton Orris Harkness, Sr., who is buried in the West Rockport Cemetery, married Ethel Frederica Packard in Maine in 1924. Their son Vinton Orris Harkness, Jr. was born in 1928.]

Two years after the death of Anna Eliza, O.A. Harkness married Alice H. Kettell (1897-1980) of Brewer, Maine, on October 18, 1930, and their daughters were Avis and Betty Alice. Avis Woster Harkness Black was born April 5, 1940, and died at age 71 on August 15, 2011 in a Bangor hospital. Her sister Betty Alice Harkness Peters currently lives in Falmouth, Massachusetts.

O.A. Harkness – Obituary

From his obituary in *The Camden Herald*, April 12, 1951: "O.A. Harkness died April 8, 1951, in Veazie, Maine, at the age of 81 after a long illness. Before his retirement in October 1950, he had been employed by GNP Co. for 36 years as Mechanical Superintendent of the Great Northern Paper Company's Spruce Woods Department. He had been a resident of Veazie for 32 years prior to his death and had retired to his home there. He was a thirty-second degree Mason and belonged to St. Andrew's Lodge, F and AM, Mount Moriah Chapter, St. John's Commandery, and Anah Temple, Order of the Mystic Shrine. He attended the Veazie Congregational Church.

Anna Eliza Fernald, the first Mrs. O.A. Harkness, died at age 65 on June 29, 1928, from burns and shock after a fire which partially destroyed her home in Veazie, Maine, near Bangor. She was survived by her husband and her two children, her daughter Elizabeth A. Harkness and her son Vinton Orris Harkness.

He was survived by his wife, Alice, and one son, Vinton, [Sr.] of Waban, MA; three daughters, Elizabeth A. Harkness of Lincolnville, and Betty Alice and Avis W. Harkness, both of Veazie, one sister, Miss Georgie A. Harkness of Rockland, and one grandson, Vinton Orris Harkness, Jr. who was in the U.S. Navy."

O.A. Harkness Early History and Experience

O.A. Harkness's first job as a youngster was helping on a naphtha steam launch, the *Titwillow* that towed the passenger barge *Mikado*, taking parties out on Megunticook Lake in Camden/Lincolnville, Maine.

Naphtha launch *Titwillow* towing the passenger barge *Mikado* on Megunticook Lake, Camden, Maine, with the Turnpike (Rt. 52) in background ca. 1887
Camden Public Library, Walsh History Center, Barbara F. Dyer Collection

The names of the steam launch and passenger barge apparently came from Gilbert & Sullivan's comic opera The Mikado which opened in London in 1885 and included a popular song about the titwillow bird (which looks similar to a black-capped chickadee).

By the 1880s, the steam engine was well established internationally as a power unit for small steam launches as well as for large ships. U.S. law, however, prompted by some boiler explosions, required that all steam boats, no matter what their size, carry a licensed engineer at all times. Although this was no difficulty for a commercial craft, becoming a licensed steam engineer required an apprenticeship of two years beforehand and effectively prevented small steam launches from being used for personal and recreational purposes. To avoid the licensing requirement, external combustion naphtha engines were used instead to power small vessels. A naphtha engine uses naphtha, a volatile and highly flammable hydrocarbon, which is derived from coal tar or petroleum, for fuel. The typical naphtha launches were about 24' in length.

In 1896 O.A. Harkness built the 41-foot steam launch *RAY.* on the Harkness family's Lincolnville property on Fernald's Neck in the middle of Megunticook Lake and launched the vessel into the lake. It was powered by a steam engine with a wood-burning boiler. By then, at about age 28, Harkness had become a licensed steam engineer, and he was the *RAY.*'s captain as well.

The 41 foot steamer, "Ray" built in 1897 at Lincolnville, Maine on the shore of Lake Megunticook and served there for excursions on the lake for about ten years. O.A. HARKNESS AT HELM.

Steam Launch *RAY.* with O.A. Harkness at the helm as captain and engineer
V.O. Harkness, Jr. Collection

This announcement offering tours aboard the boat was placed in
The Camden Herald
V.O. Harkness, Jr. Collection

An 1897 ad described the location of the cruise waiting room and wharf as being the Harkness cottage at Lake City on the Turnpike [Route 52] side [probably near the current Barret's Cove]. The *RAY.* carried passengers on Megunticook Lake from 1897 until about 1905, when she was sold and moved to Aroostook County where she was used on Portage Lake.

Education

No historical record or family recollection has been able to provide information concerning O.A. Harkness's education. V.O. Harkness, Jr., recalls, however, that the family lived in the Rockport section of Camden (the two Towns did not separate until 1891) and probably attended Camden High School in the 1883-87 time period.

Employment and Work Experience, pre-1901

At some point between the late 1880s and before 1901, Harkness was employed in the experimental department at the Walworth Manufacturing Company in Boston and at Bath Iron Works in Bath, Maine. The author's research has not been able to discover the specific years he worked for either one of them. Harkness's years at these two employers would be bracketed, however, by his presumed 1887 high school graduation and his employment with Eastern Manufacturing in Brewer, Maine, where he began work as Master Mechanic in 1901. Brief histories of these two companies are provided below to give readers a sense of the kinds of knowledge and work experience he might have gained during that time period.

Also, it appears he was self-employed in 1896 when he built the steam launch *RAY.* in Lincolnville and captained that vessel at least during the summers of 1897-98.

Walworth Manufacturing Company

In 1841, J. J. Walworth and Joseph Nason formed a partnership in New York, under the firm name of Walworth & Nason, for the purpose of providing heat and ventilation to buildings by means of steam and hot water. The company was originally engaged in contracting and construction work, providing necessary supplies and installing heating systems in homes and businesses. It used hot water as a means of heating until 1844 when it began to utilize steam. The company moved to Boston in 1842 and established a small plant in the heart of the city. Five years later the plant was transferred to Edgeworth, Pennsylvania, but an office was maintained in Boston. The plant was employed to manufacture pipe which was used by the company in its installation work, although the manufacturing operations of the company continued to be secondary to its contracting work. In 1852 Joseph Nason withdrew from the partnership, and a new partnership was formed under the name of J. J. Walworth & Company. This partnership was incorporated in 1872 under the name of the Walworth Manufacturing Company.

In the late 1850s the company commenced the manufacture of pipe fittings and later it undertook the production of valves. These were manufactured only because the company found it difficult to purchase elsewhere the fittings and valves which it needed in its construction work. As its activities increased, the company found its original plant insufficient and a new plant was established at Cambridgeport, Masachusetts, about 1860. This plant was continued in operation until 1882, when a new factory was constructed in South Boston. The company continued to produce pipe fittings and valves and carry on its construction work until 1912. In 1922 the company ceased to do construction work and to produce pipe, devoting its entire energy to the production of fittings and valves. [4]

Bath Iron Works

A brief description of the ships built at Bath Iron Works (BIW) in the 1888-1901 time period describes some of the ship construction knowledge and experience Harkness could have gained from working there.

From Ralph Linwood Snow's *Bath Iron Works The first 100 Years*, Appendix A, Bath Iron Works Hull List:1888-1987, Maine Maritime Museum, Bath, Maine, 1987, p. 290 and p. 574-576, BIW was founded in 1888 and initially built large wooden-hulled ships with steam engines:

- 1890-91 two passenger ships, 232' in length and 40' beam, with 1300 hp steam engines
- 1890-91 two U.S. navy gunboats, 190' x 32' with 2046 hp steam engines
- 1891 a U.S. Navy ram 250' x 43'
- 1892-93 a passenger ship, 322' x 50' with a 4500 hp steam engine
- 1893 a bark 136' x 30' with a 500 hp steam engine
- 1896-97 a tugboat, 151' x 16' with a 4200 hp steam engine
- 1897 two lighthouse vessels 113' x 30'
- 1898 three tugboats 157' x 17'
- 1899 a U.S. Navy light cruiser 292' x 43', 3200 tons with a 4460 hp steam engine

Additionally, the General Dynamics Corp.'s website for BIW lists two steam yachts built in 1896: the 136' *Peregrine* for R.H. White of Boston, and the 110' *Illawarra* for Eugene Tompkins of Boston. These two vessels are similar in size to the 104' steam yacht *Helena* for which Harkness took charge of construction at the Eastern Manufacturing Company in Brewer, Maine, in 1901 (see the Eastern Manufacturing Company information which follows).

Eastern Manufacturing Company

Frederick Wellington Ayer (F.W. Ayer) bought the large Palmer and Johnson sawmill in South Brewer in 1881, and in the mid-1880s installed the first bandsaw in eastern Maine. It was the "wonder of the age" for its productivity. In 1899 Ayer began manufacturing paper (using the newly-developed sulfite process which he patented) to utilize the leftover slab lumber from the sawmill. That new method of sulfite digestion subsequently became the basis for operations at the Eastern Manufacturing Company, as well as for the Great Northern Paper mill in Millinocket, Maine. By 1912 Eastern Manufacturing reportedly employed 1,000 workers at its paper mills and had a crew of 1,000 cutting timber.

1921 image of the Eastern Manufacturing Company, Brewer, Maine
www.mainememory.net

O.A. Harkness was hired as Master Mechanic for the operations of the Eastern Manufacturing Company and worked in Brewer from 1901 to about 1903. In November 1901 one of his first tasks was to take charge of the construction of the steam yacht *Helena* being built for company owner F.W. Ayer.

The *New York Times* reported the launching on April 1,1903, and the vessel was described by the *Marine Review and Marine Record* in its July 1903 edition as follows: "A special undertaking in ship building ways at this place [Brewer, Maine] was the building of a 104' steam yacht for F.W. Ayer, costing $50,000, almost entirely from the facilities of the paper mill plant of the Eastern Manufacturing Co. at South Brewer. The craft was built of hard pine and her engines were made in the machine shops of the Company. There has been a brisk business in small lake steamers and naphtha launches, and about a dozen have been turned-out by Brewer boat builders during the past two years."

Helena is listed as vessel No. 316 in the 1906 edition of *Lloyd's Register of American Yachts.*

Named for Ayer's daughter, the 104' vessel was designed by Eastern Manufacturing executive Charles B. Clark and had a beam width of 17' 3", weighed 94 gross tons, had a mahogany deck house, and was schooner rigged. [5]

Steam yachts often carried sails in those days as a precaution in case the engines failed. The vessel was sold by Ayer in 1907.

Although no photographs of the *Helena* have been found by the author, a similar-sized vessel, the *Machigonne,* was built in 1903-04 for publisher Cyrus H. Curtis, whose summer home was in Rockport, Maine. *Machigonne* was 119' in length with a beam of 16' 6", and was schooner rigged.

Steam yacht *Machigonne*, whose summer home port was Camden harbor
The Motor Boat magazine, Vol 4, No. 3, February 10, 1907, p. 51

The 1907 U.S. Census shows O.A. Harkness as a resident of Lincolnville with his wife and two children and lists his occupation as boatbuilder.

Penobscot Log Driving Company

Each spring the logs harvested by the various logging companies in the upper Penobscot River basin were driven down the river to the "booms" in the lower Penobscot, and timber from dozens of companies was floating down the river at any given time. In order to avoid conflict, timberland owners formed mutual benefit companies to coordinate driving, sorting, and rafting activities.

One of these companies was the Penobscot Log Driving Company (PLDC), chartered by the state legislature in 1846. The PLDC initially earned approval through the legislature to build dams, blast rocks, and improve Penobscot River channels for improving "navigation." Those improvements resulted in regulated lake water levels which allowed snowmelt water to move logs to the mills in springtime log drives.

In 1903 Harkness was assigned by the PLDC to build a steam boat to tow booms across Chamberlain Lake on the East Branch of the Penobscot River, and he left Bangor on January 1, 1903, to undertake that project. That Harkness-built steamer, named the *George A. Dugan*, was 71 feet long with a 20-foot beam, drew four feet of water, and was powered by two steam boilers and two engines. It was completed in May 1903 and was kept in operation for about 10 years.

In 1903 (Chapter 127 of State Statutes) the Maine legislature gave control of the waters of the West Branch above Shad Pond at Millinocket to the West Branch Driving and Reservoir Dam Company, a GNP Co. subsidiary, replacing the PLDC.

The log drives on the West Branch of the Penobscot River initially floated 20-foot or 24-foot "long logs" to the saw mills, and the last long log drive took place in 1928. Pulpwood for making paper was cut into shorter 4-foot lengths beginning in 1917. Over time, the transport systems for harvested timber evolved from horses and oxen, to a tramway, to Lombard log haulers, to a railroad, to crawler tractors, and eventually to wheeled vehicles with pneumatic tires, all to haul logs from the East Branch, West Branch, Allagash, and St. John River basins to the West and East Branches for transporting them via towboats and dam systems to the paper mills on the Penobscot River.

The Great Northern Paper Company

O.A. Harkness left the Eastern Manufacturing Company in 1915 and was then employed with the Great Northern Paper Company until about 1951. He served first as Superintendent of Motor Equipment and later as Superintendent of Mechanical Equipment.

By 1915 Harkness's experience and skills in boatbuilding and the knowledge of motor-driven equipment of all kinds were well developed through his prior employment history. Those skills, combined with his innate talent for mechanical engineering, his creative and practical focus on operational efficiency, his ability for strategic planning, and his seemingly boundless personal energy, made him a dynamic and important figure in the transport of northern Maine timber.

Chapter 2

A Scow, and Early Side-Paddle Wheel Tow Boats

Towing Log Booms on Lakes

A log boom (sometimes also called a boom bag) was a barrier placed in a lake or river, designed to collect and/or contain floating logs timbered from nearby forests. The log boom itself was constructed of a series of 28'-30' logs fastened end-to-end by log booms chains and formed into a circle with an open mouth. Logs were floated into smaller booms of about 5 acres in size which were then towed across lakes by 30'-35' boats. They were then gathered into "trip" booms as large as 20 acres in size to be towed across lakes, like rafts, and then sluiced by the force of river currents to saw mills and paper mills. Booms were also anchored in rivers to contain logs while they awaited their turn to be processed at the mills. Booms prevented the escape of these valuable assets into open waters.

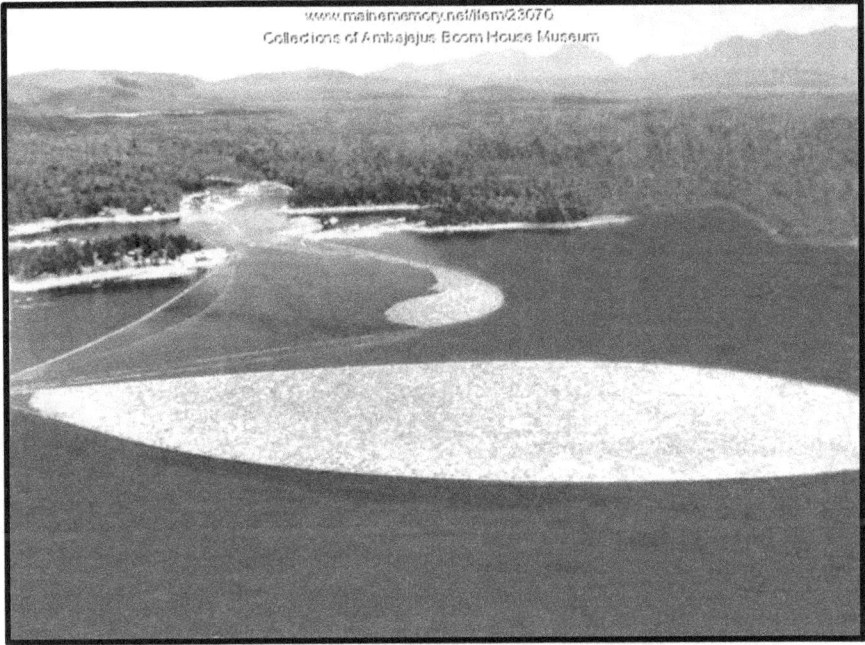

Two log booms "bags" on Ambejejus Lake near Millinocket
The larger boom is a "trip" bag ready for towing, and the
smaller one is only partially filled.
Maine Memory Network, Collection of the Ambejejus Boom House Museum

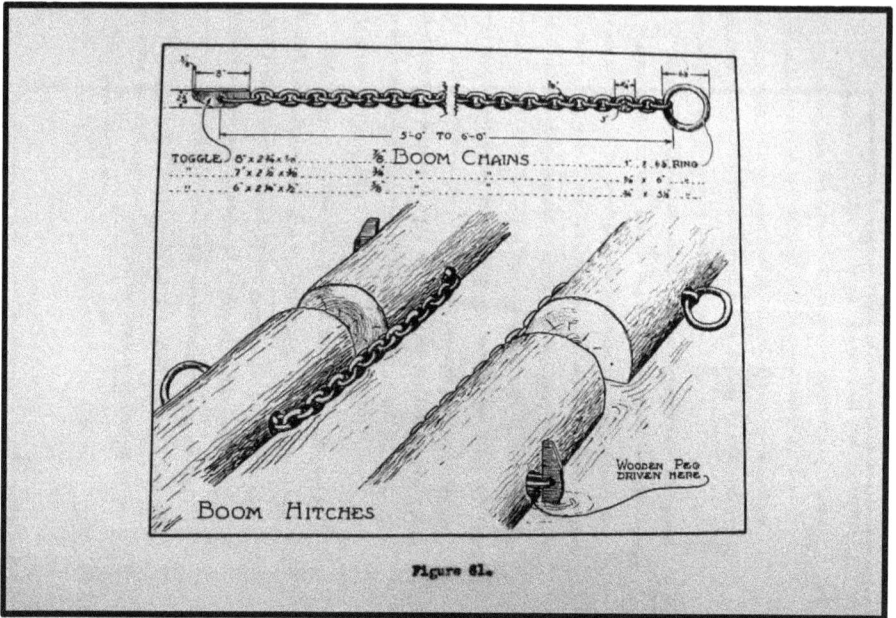

Boom chain detail
Cecil Max Hilton, Equipment and Methods Used at Pulpwood Operations
Upon the West Branch of the Penobscot River in Maine, 1935-40,
University of Maine Thesis, 1940. Great Northern Paper Company Records,
SpC MS 0210, Box 13, Folder 1, Raymond H. Fogler Library Special
Collections Department, University of Maine, Orono, Maine

The Penobscot Log Driving Company (PLDC) charged each "operator" (sawmill or paper mill owner) a fee for the number of board-feet put through to the river booms each season. In the lower Penobscot near Old Town, Orono, Bangor, and Brewer, the log drivers sorted the logs into fixed booms by each operator's "mark" cut into the log at the time it was harvested. PLDC scalers then measured the lumber in the booms to calculate each operator's fee for having driven the logs down-river.

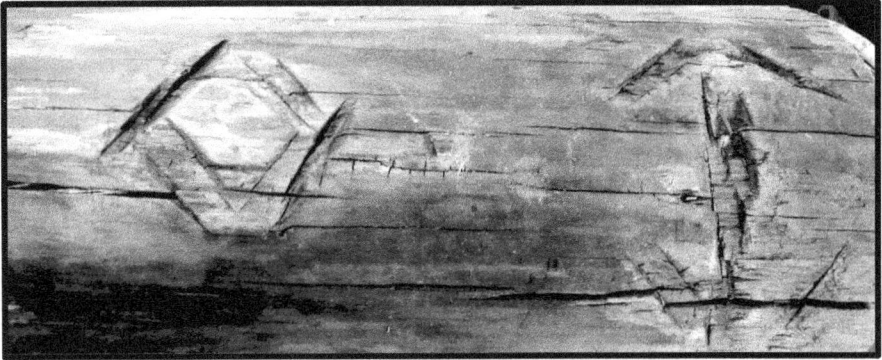

**"Diamond" and "double dart" operator's marks cut into a log
at the Patten Lumbermen's Museum, Patten, Maine.**
Photo by the author 2017

Before side-paddle steam tow boats came into use, moving log booms down a lake involved incredible labor. The booms were towed by a "headworks," a heavy raft on which was mounted a primitive capstan, made from an upright log, which pulled the log boom cable. To move the log boom, the anchor was boated-out ahead by a small rowboat to the length of the boom cable and dropped overboard. Then a team of ten to twelve strong men manned the capstan bars and walked the raft, towing the boom up to the anchor. The anchor was lifted and the process repeated. Two or three miles was all that could be gained in twelve hours of toil. [6]

In favorable weather, crews sometimes worked for three days and nights to move a boom down the lake. This was hard, slow, and expensive work. [7]

Capstan and crew
Collection of Patten Lumbermen's Museum, Patten, Maine

Early Side-Paddle Wheel Tow Boats

The first log boom tow boat on Chesuncook Lake (the third largest lake in Maine) was the steam side-wheeler scow *John Ross* (owned by the PLDC and named for its president). In 1890 when the *John Ross* was built, there was no road up the east side of Moosehead Lake to Chesuncook Lake. Building materials for the vessel were taken up Moosehead Lake, and the *John Ross* was built at that lake's Northeast Carry. When finished, the *John Ross* was hauled across four miles of land to the West Branch of the Penobscot River where it was run down over the West Branch's Pine Stream Falls to Chesuncook Lake by Captain Louis Gill. At the falls, it had to be let down on a line with blocks, tackles, chains, and pulleys, and when the right point was reached, Captain Gill gave the signal and the line was cut. The boat ran the rest of the whitewater without engine power. Once in operation, the two 25 horsepower steam boiler engines proved to be so heavy that it was feared they would drop through the scow's bottom, so stanchions were built in the hull from which chains were hung to hold the weight of the machinery. [8]

Because the *John Ross* was designed as a scow, however, it didn't steer well at first because of its rudimentary rudder. A false prow was added which improved steering, but the prow leaked badly and was always full of water. In its first booms the *Ross* averaged only about two million board feet per boom, but the last boom of the year was four million feet. That last boom proved too much for the *Ross* to handle, however, and the main paddle shaft broke under the strain. Repairs were made and the *Ross* ran for about 11 years until 1901 when it was scrapped. It was replaced in 1902 by the newly-constructed *A.B. Smith* (named for Ansel B. Smith), also a side-paddle wheeler. The *John Ross*'s engines had survived the boat itself, though, and were placed in the *A.B. Smith*. After replacing the *Ross,* the *A.B. Smith* towed log booms on Chesuncook Lake until about 1927. [9]

A.B. Smith
Chesuncook Village Association website photo collection

A.B. Smith
Moosehead Historical Society Collection

Note: Both boats depicted have the name *A.B. Smith* clearly marked on the side paddle structure and wheelhouse roof respectively, but the superstructures are different and must have been modified at some point.

The use of tow boats for pulling log booms across lakes had proved so successful that in 1892 the PLDC built the side-paddle steamer *F.W. Ayer* at North Twin Dam to use on the "lower lakes" (Ambajejus, Pemadumcook, and North and South Twin Lakes) near Millinocket. It was launched in 1893 and operated until it was replaced by the *West Branch No. 1* in the early 1920s.

The *Ayer* was captained by Sam Bordeau, a French Canadian who had become a United States citizen. [10]

F.W. Ayer with dory
www.mainememory.net photo from the Norcross Heritage Trust Collection

Towing pulpwood log booms across the open lakes was relatively simple, although the tow boats were limited to operating at one mile per hour. A higher speed would cause the 4' pulpwood logs to jump over the back edge of the containment boom. Maneuvering booms through narrow lake channels could be tricky and limited the number of logs that could be towed at one time. Wind storms could cause an entire tow to be abandoned until the weather moderated.

The H. W. Marsh, built at end of Tramway, Eagle Lake, and used there five seasons

H.W. Marsh under construction at Eagle Lake in 1903
The Northern, November 1927, pg. 15. Great Northern Paper Company
Records, SpC MS 0210, Box 21, Raymond H. Fogler Library Special Collections
Department, University of Maine, Orono, Maine

The H. W. Marsh completed her service on Chamberlain Lake about 1913

H. W. Marsh

The Northern, November 1927, p. 15. Great Northern Paper Company Records, SpC MS 0210, Box 21, Raymond H. Fogler Library Special Collections Department, University of Maine, Orono, Maine

The side-paddle steamer *H.W. March* (named for the lumber baron Herbert W. Marsh) was built on Eagle Lake about 1903. It was 91' long, had a 25' beam, and drew 4' of water. The steamer was powered by a 150 hp cross compound steam engine with two vertical wood burning boilers.

It was used on Eagle Lake for about five years to tow log booms to the tramway (see Chapter 7), and she was then brought over to Chamberlain Lake to assist the *George A. Dugan* in towing booms to the East Branch of the Penobscot.

Although the bow was normally safely pulled up on the shore, in about 1913 the stern became partially frozen into the lake. To protect the engine the stern was cut off just behind the paddle box, and when the ice went out in the spring, the stern floated away and was lost. The useful life of the *H.W. Marsh* thus ended after ten seasons with its being cut in two on the shore at Chamberlain Farm. [11]

These four side-paddle boats (*Ross, Smith, Ayer,* and *Marsh*) were neither designed nor constructed by O.A. Harkness, although he likely had operational control over them after 1903.

Chapter 3

Propeller-Driven Tugs

On January 1, 1903, Harkness left Bangor to design and build a new steam boat for the PLDC to tow booms across Chamberlain Lake as part of a plan to move the logs into the East Branch of the Penobscot. This new tow boat, the *George A. Dugan* was to be 71 feet long with a 20-foot beam. It was to be powered by two steam boilers and two engines and to draw four feet of water. [George A. Dugan was a management employee at the Eastern Manufacturing Company in Brewer.] The *Dugan* was built at Chamberlain Farm from timber cut nearby, and construction was completed in the amazingly brief time of four months. It was launched in May 1903, and the vessel was kept in operation for about 10 years. It was twin-screwed and powered by two single engines and two vertical boilers, capable of delivering 150 horsepower. [12]

George A. Dugan **under construction at Chamberlain Lake**
GNP Co. photo from the Maine Forest and Logging Museum

Tow boat *George A. Dugan*
Collection of model boat maker Peter Templeton, Greenville, Maine

The lumber and paper mill owners learned that with using tow boats in combination with systems of dams, log drives could begin when the mill operators wanted the timber without waiting for spring snowmelt when natural water levels were high enough to float logs to the mills. Furthermore, the dams were increasingly built with the additional purpose of generating electricity to power the paper mills, thus increasing the mills' economic efficiency and competitive advantage. With ship propulsion technology changing, more propeller-driven tow boats were on the way.

GNP Co. *No. 21*

"Boat *No. 21* marked an epoch in the annals of the country in the heart of the Maine Woods and a new era in the work of the Great Northern Paper Co. – more especially in the progress of the Motor Boat Division headed by O.A. Harkness." Harkness designed boat GNP Co. *No. 21* and supervised its construction by a crew of expert boat builders.

Charles "C.H." Ingalls, a master boat builder from East Machias, Maine, built a boat model from the Harkness plans and then directed the vessel's construction. It was launched on May 24, 1921, into Chesuncook Lake where it was to be based. Fifty-eight feet in length and with a beam of 14 feet, it drew 5½ feet of water. It was powered by a 60 horsepower Fairbanks-Morse diesel engine which drove a 48" diameter propeller. It also had a 600 candle-power search light for nighttime boom towing operations. After each springtime log drive was finished, it was put to work carrying passengers and freight on Chesuncook Lake. With two cabins, one fore and one aft, it could comfortably accommodate nearly 50 passengers. [13]

Note: O.A. Harkness's grandson, Vinton Orris "V.O." Harkness, Jr., provided information that his father Vinton Orris Harkness, Sr. had worked on GNP Co. *No. 21* while he was a student in mechanical engineering at the University of Maine. He graduated in 1922, and his working career was at Fairbanks-Morse for 42 years.

Its year of construction, lake location, size, and configuration with a diesel stack and searchlight all point to the vessel in the photo below as being GNP Co. *No. 21*. The name over the starboard (right) side of the pilot house clearly says GNP Co., but the resolution of the photo isn't clear enough to discern the lettering on the port (left) side of the pilot house.

The author presumes that the boat in this photo is GNP Co. *No. 21*.
Stephen Rainsford Collection, Bowerbank, Maine, courtesy of Ellen Rainsford.

GNP Co. *West Branch No. 1*

Except for this photo, very little information is provided in GNP Co. records about *West Branch No. 1*. It replaced the *F.W. Ayer* on the "lower lakes" near Millinocket in the early 1920s after Harkness was employed by GNP Co. and was in service until 1942 when it was replaced by the *West Branch No. 3*. The vessel was initially steam powered but was later converted to diesel.[14]

West Branch No. 1, **note the dory towed alongside.**
The Northern, September 1922, p. 9 Great Northern Paper Company Records, SpC
MS 0210 Box 21, Raymond H. Fogler Library Special Collections Department,
University of Maine, Orono, Maine

GNP Co. *West Branch No. 2*

By 1926 the *A.B. Smith,* built in 1902, needed replacement, and O.A. Harkness created the model for the *West Branch No. 2* as its successor. He prepared all the specifications and directed H.W. Wright in drawing the plans. There was no shipyard at hand, so Great Northern created a shop in Greenville to prepare the wooden building materials for the boat's construction. The shop was equipped with planers, saws, drilling machine, and an emery wheel, all powered by a gasoline engine. A boiler and steam box were set up for steaming the planking; steaming boat ribs and planks makes them flexible enough for bending to meet the designed shape of a vessel's hull.

Most of Great Northern's employees were then unfamiliar with boat construction, and when the building crew was assembled there was scarcely a man at the project who had ever built a boat before, but building the *West Branch No. 2* was their chance to learn. The exception was the master builder Charlie "C.H." Ingalls, a master boat builder from coastal East Machias, Maine, who led the construction of the vessel.

The building materials for the boat included 17,000 feet of oak for the ribs, and a larger quantity of hard pine for planking. The 12" x 12" hard pine keel was laid in May 1926, with a 6" birch shoe with an iron guard over it. The strakes [hull planking next to the keel] were 5" thick, and the hull planking was 3" thick.

The *West Branch No. 2* was actually built in a shipyard at the Chesuncook Lake Dam from the hull and cabin construction materials prepared in Greenville. The vessel was fitted with an engine house, dining room, galley, and staterooms for crew and guests. The forecastle berthed six men. The vessel was powered by a 360 horsepower Fairbanks-Morse diesel engine which weighed thirty tons. It came to Greenville by train, then to Lily Bay on Moosehead Lake by scow, and finally to the dam by trailer. The vessel had five one-thousand-gallon fuel tanks distributed from bow to stern for balance, enough for ten or twelve days hauling, even for around-the-clock work. Another small diesel engine was aboard to generate electricity for lights, searchlights, and an electric winch for making the work smoother.

The West Branch No. 2 was launched at Chesuncook Dam May 11, 1927

West Branch No. 2
The Northern, June 1927, pg. 4. Great Northern Paper Company Records, SpC MS
0210, Box 21, Raymond H. Fogler Library Special Collections Department,
University of Maine, Orono, Maine

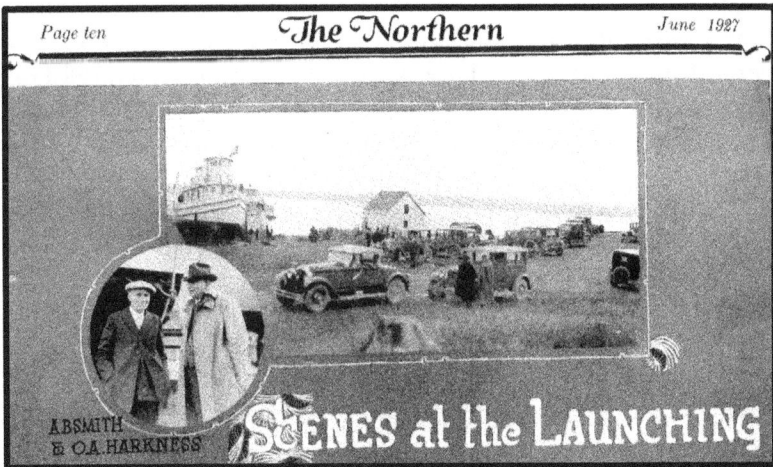

A.B. Smith and O.A. Harkness at the launching of the *West Branch No.2*
The Northern June 1927, pg. 10. Great Northern Paper Company Records, SpC MS
0210, Box 21, Raymond H. Fogler Library Special Collections Department,
University of Maine, Orono, Maine

The new craft was an instant success as the following statistical comparison of the two boats by O.A. Harkness will readily show.

	A.B. Smith **1926**	**W.B. No. 2** **1927**
Started towing	May 19	May 21
Finished towing	Sept. 4	July 21
Time towing	3 mos. 15 days	2 mos. 6 days
Crew	10 men	7 men
Time taking fuel	100 hours	13 ½ hours
Fuel used per 24 hours	10 tons of coal	301 gallons
Fuel cost per day	$250	$50
Number of booms towed	62	51

In towing the 51 booms, there were 13 days saved on the time it took to tow the 51 booms with the *A.B. Smith* the year before. [15.]

It was said that tugs like the *West Branch No. 2* could tow booms of about 21 acres in size. [16]

A typical tow, however, seems to have been 4,000 to 5,000 cords. At the end of its useful life of about 34 years, *West Branch No. 2*, the once proud monarch of the lake, was towed to a cove called Holmes' Hole in the early 1960s and burned. [17]

Its replacement in 1961 was the steel heavy tug *William Hilton*.

***West Branch No. 2* towing a log boom on Chesuncook Lake**
Great Northern Paper Company Records, Raymond H. Fogler Library
Special Collections Department, University of Maine, Orono, Maine

GNP Co. *West Branch No. 3*

In 1942 O.A. Harkness made the model for the *West Branch No. 3* and supervised the vessel's construction. Its Master Builder was Howard Watts of Roque Bluffs in Washington County, Maine. The boat was 91 feet long with a beam of 21 feet and was equipped with a 200 hp Fairbanks-Morse diesel engine. It had a gas refrigerator and cooking range, and the galley was equipped with hot and cold running water. A generator provided lighting, and there were sleeping accommodations for eight. [18] Its Captain was Bob Sawyer.

West Branch No. 3 **under construction, July 1942**
V.O. Harkness, Jr. Family Photo Collection

West Branch No. 3 **under construction, with O.A. Harkness on deck, July 1942**
V.O. Harkness, Jr. Family Photo Collection

This GNP Co. photo shows the *West Branch No. 3* just after its launching at North Twin Lake in the "lower lakes" in May 1943
Great Northern Paper Company Records, SpC MS 0210, Raymond H. Fogler Library
Special Collections Department, University of Maine, Orono, Maine

Lew Dietz wrote the following caption for this photo in the *National Fisherman,* May 1967, page 18-A, in an article entitled *Special Breed of Boats Worked Maine's Lakes.* "***WEST BRANCH NO. 3*** was among the handsomest vessels in the Harkness 'navy.' The hull was designed by Harkness himself, although McLeod [John E. McLeod, GNP Co. officer and author] believes that the house came from a vessel on Moosehead Lake. She worked the lower lakes until replaced in the '60s" [replaced by the steel-hulled boom tug *O.A. Harkness* in 1964].

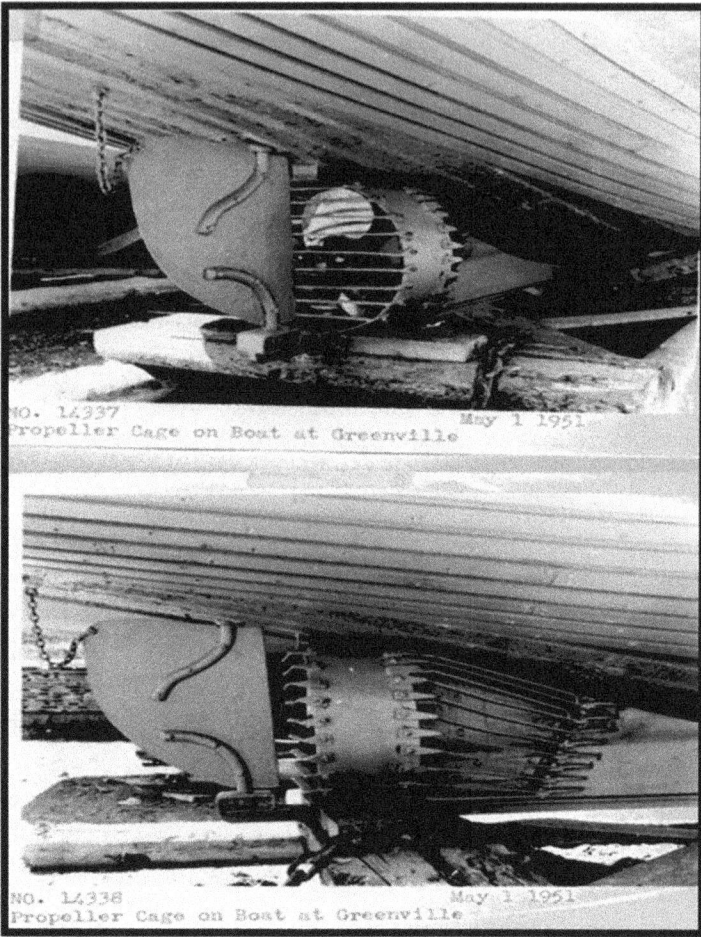

Tow boat propeller cages (1951) protected propellers from damage
by pulpwood logs
Chesuncook Village Association website photo

Chapter 4

Boom Jumpers

O.A. Harkness, on the suggestion of Great Northern's Vice President Fred A. Gilbert, developed the smaller "boom jumper" boats to fill the need for a boat that could operate in water filled with four-foot long pulpwood held together by floating boom logs which needed to be crossed. The boom jumpers were used "primarily to do errands and light work around booming-out places and in bringing up and placing the slack boom. When it is necessary to cross a boom and get inside, the boom jumper is headed directly upon it. The engineer opens the throttle and the bow hits the boom with a splash; there is a moment's hesitation, then the boat rides smoothly on the forefoot and slides easily over." [19]

Boom jumpers were also useful for towing small log booms in "sweeping" operations, where pulpwood remaining in flat water areas after a main body of logs had passed through a dam was gathered into small booms towed by boom jumpers. These boats also saved many miles of travel in lakes where many booms were required or where working in such pulpwood-filled waters would have required long detours in ordinary boats. [20]

About 1914 Fred Sawyer of Greenville (Maine) then employed by the GNP Co. built *No. 10*, the first of the flat bottom, three-keel type. It was rebuilt in 1923 at the Greenville shop, and heavier planking was used.

When Harkness became employed by the GNP in 1915, there were seven motor boats in use. Five were Atlantic dories with Atlantic engines with horsepower ranging from 8 to 15. [Atlantic dories and Atlantic engines were built at the Lunenburg Foundry, established in 1891 in Lunenburg, Nova Scotia. Its famous Atlantic Marine engines, whose signature *putt-putt-putt* sound can still be occasionally heard barking in coastal waters, were called "one-lungers" and "make-and-breaks."]

Note: The only image the author has been able to find of an Atlantic Dory with an Atlantic Engine is shown here.

Atlantic Dory on Lake Chatauqua, N. Y.

DORY ON INLAND WATERS

THE above photo is of one of the Atlantic Company's 30-ft. semi-speed clipper launches, equipped with Atlantic Special 15-18-h.p. engine. The boat is used on Lake Chautauqua, and is owned by Mark Packard, of Buffalo, N. Y. The craft was designed to carry a large party, as may be seen from the photo, with seats around the sides of the cockpit, engine installed in the forward cockpit, where there are side seats, control of the engine through reverse gear, and the forward control handles.

Although of the regulation coast type the craft has given excellent satisfaction, and should prove a comfortable and safe craft on any inland body of water, particularly localities subject to sudden squalls and rough water.

Atlantic dory with an Atlantic engine
The Rudder magazine, July-December 1910, edited by Thomas Fleming Day, New York, New York, Vol. 24, p. 315

Although the dory in this photo is adapted for carrying passengers, it seems likely that this 30' craft served GNP Co.'s early needs for lake transportation. Its appearance is similar to the plan for boats *No. 17* and *No. 18* on the following page.

No. 3 and *No. 10* were the other two boats. *No. 3,* a round bottom boat, was built in Camden by the Camden Anchor and Rockland Machine Company about 1912. It was 37' long and had a four-cylinder, 4-cycle 40-hp [gasoline] engine built by the same company. It was rebuilt in 1922 when a semi-diesel engine was installed.

Harkness designed *No. 11* and had it built in Brewer in 1916 by the Cobb Brothers boatyard. This boat was a round bottom boom jumper with three keels. In this type of boat, the main keel is in its usual center place and there are also keels on either side of it. These keels project under the stern to make a box which guards the propeller from the floating wood, and it also protects the bottom and the wheel when the craft goes over a boom.

In 1919 two more boats, *No. 17* and *No. 18*, were constructed by the Cobb Brothers [from plans drawn by O.A. Harkness]. Both were about 30' long with flat bottoms, had three keels, and were powered by 4-cycle 4-cylinder 40-hp Lathrop gasoline engines.

Great Northern Paper Company Records, SpC MS 0210, Raymond H. Fogler Library
Special Collections Department, University of Maine, Orono, Maine

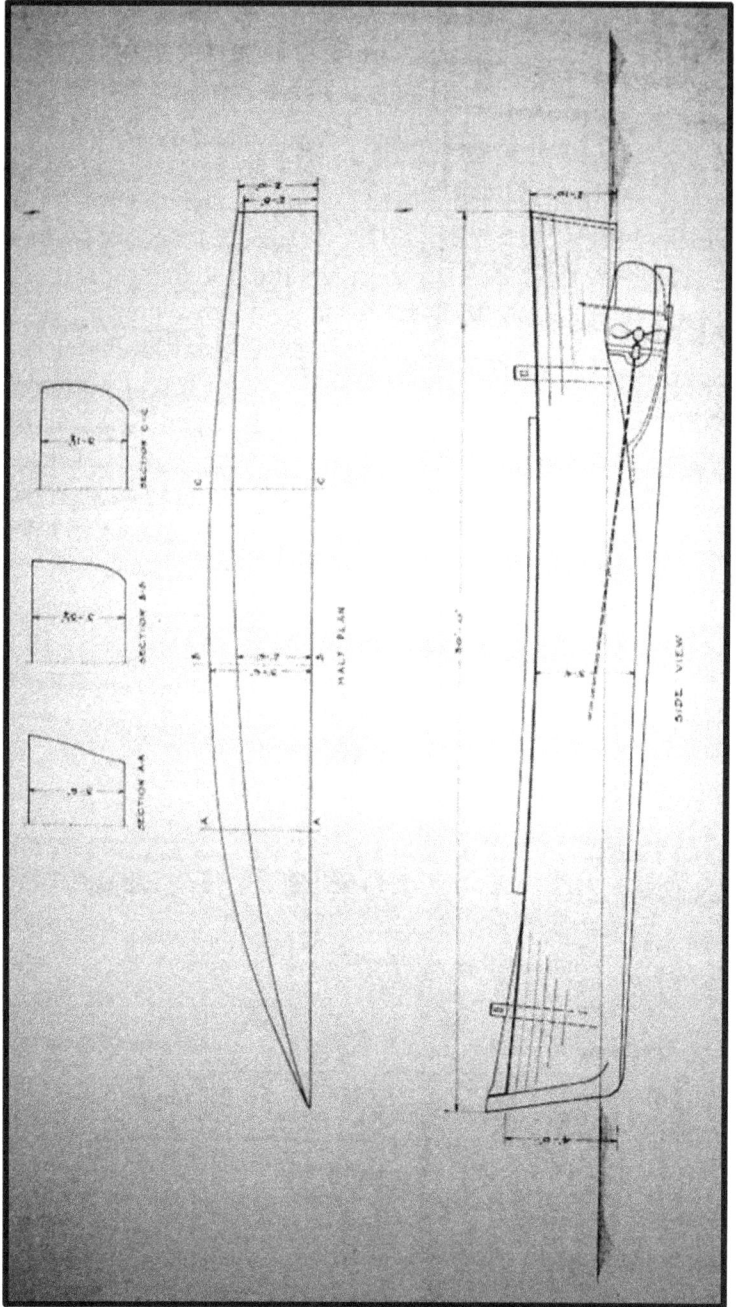

O.A. Harkness design for boom jumpers *No. 17* **and** *No. 18*
Great Northern Paper Company Record, SpC MS 0210, Raymond H. Fogler Library, Special
Collections Department, University of Maine, Orono, Maine

In 1920, *No. 19*, of the same general design, was built in Greenville by William St. Germaine, and *No. 20* was again built by the Cobb Brothers.

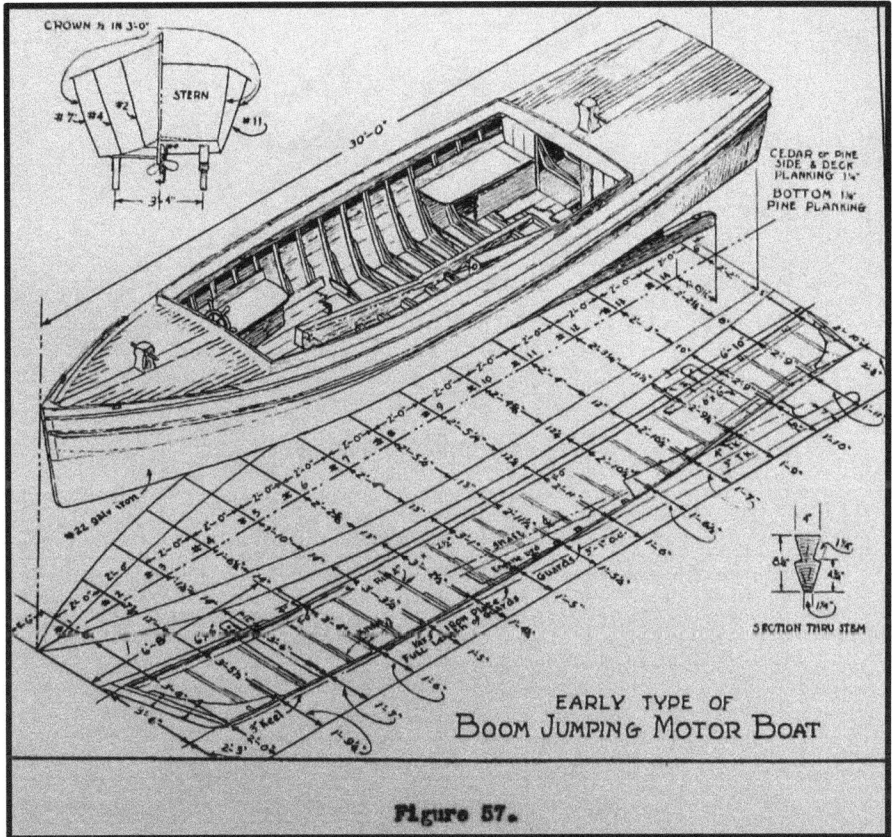

Early type of boom jumper
Cecil Max Hilton, Equipment and Methods Used at Pulpwood Operations Upon the West Branch of the Penobscot River in Maine, 1935-40, University of Maine Thesis, 1940. Great Northern Paper Company Records, SpC MS 0210, Box 13, Folder 1, Raymond H. Fogler Library Special Collections Department, University of Maine, Orono, Maine

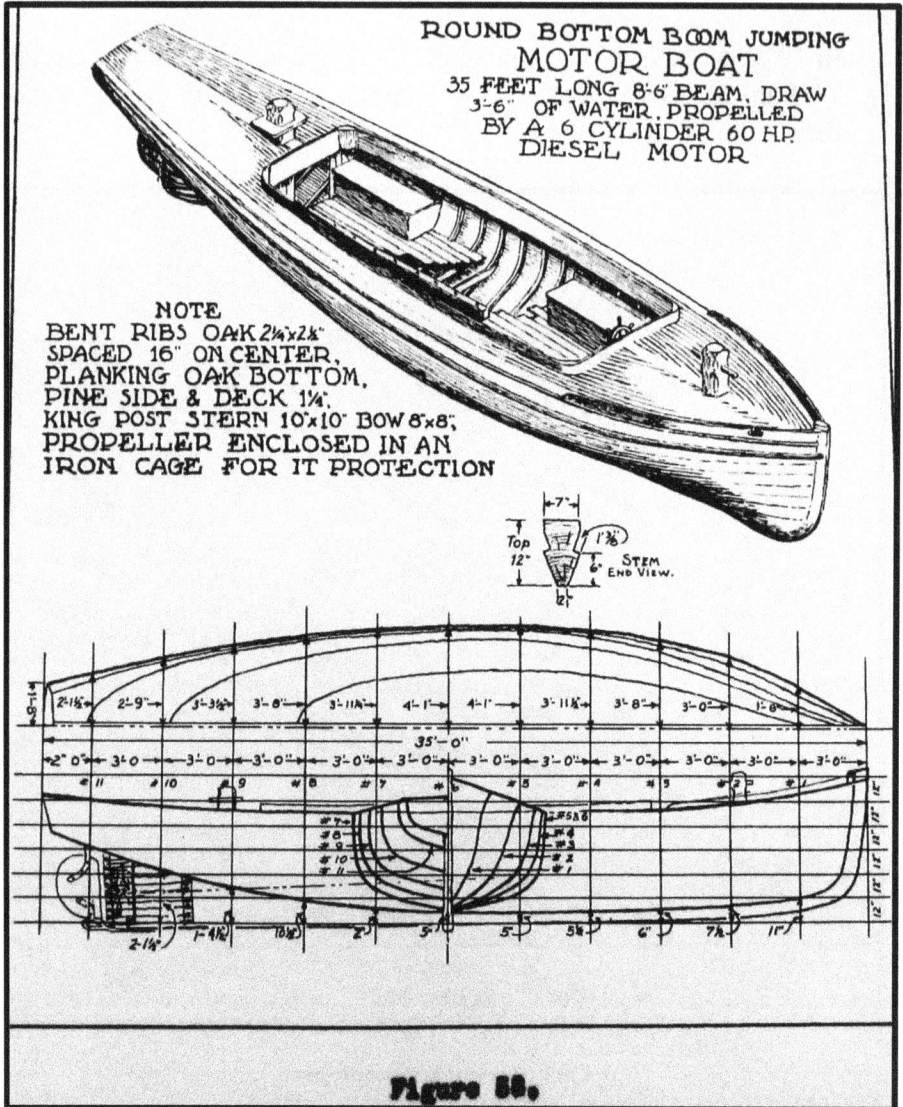

35' open-decked boom jumper with iron propeller cage
Cecil Max Hilton, Equipment and Methods Used at Pulpwood Operations Upon the West Branch of the Penobscot River in Maine, 1935-40, University of Maine Thesis, 1940. Great Northern Paper Company Records, SpC MS 0210, Box 13, Folder 1, Raymond H. Fogler Library Special Collections Department, University of Maine, Orono, Maine

Beginning in 1923, all GNP boom jumper boats were built at the company's Greenville machine shop. In 1923 *Nos. 22, 23, 24,* and *25* were built using heavier timbers and planking than *No.'s 17, 18, 19* and *20. No.'s 27, 28,* and *29* were constructed in 1924, and *No. 30* in 1925.

Thus, *No.'s 22-30* were all of the same general type.

Boom jumper keel detail
The Northern, September 1928, pg. 4
Great Northern Paper Company Records, SpC MS 0210, Box 21, Raymond H.
Fogler Library Special Collections Department, University of Maine, Orono, Maine

"When we come to Nos. 31 and 32, however, changes are to be noted. The usual length of 30 feet is increased by a foot or two. The bottoms and one streak [*sic*] of siding are of oak. The lines of the bow are changed: there is a flare to turn back the spray, a cabin is added, and a different type of engine, a 40-hp diesel is used. This, of course, burns fuel oil instead of gasoline which is a great economy as only half the amount is needed and the oil costs only half the price of gasoline. Special features of these boats are the stern bearings and the rudder construction, the patterns for which were designed by Mr. Harkness." [21]

GNP Co. *No. 31*
The Northern, July 1927. Great Northern Paper Company Records, SpC MS 0210, Box 21, Raymond H. Fogler Library Special Collections Department, University of Maine, Orono, Maine

GNP Co. *No. 36* with iron propeller cage (1930s)
Chesuncook Village Association website photo

This type of boat was developed by Mr. Harkness, following the suggestion of Vice-President F. A. Gilbert

GNP Co. *No. 37*, built by Great Northern Paper Company at its Greenville Shop
The Northern, September 1928, pg. 3
Great Northern Paper Company Records, SpC MS 0210, Box 21, Raymond H. Fogler Library Special Collections Department, University of Maine, Orono, Maine

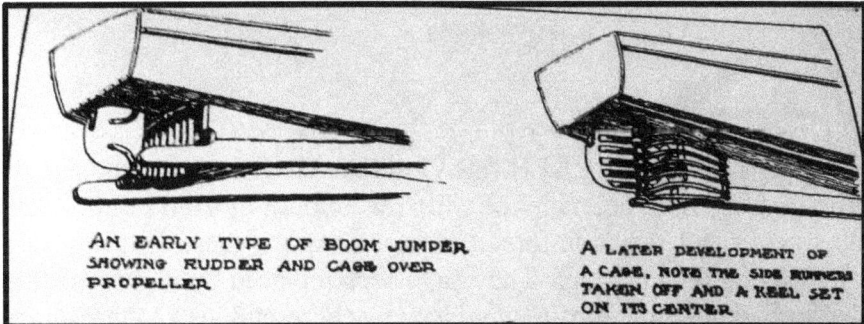

AN EARLY TYPE OF BOOM JUMPER SHOWING RUDDER AND CAGE OVER PROPELLER.

A LATER DEVELOPMENT OF A CAGE, NOTE THE SIDE RUNNERS TAKEN OFF AND A KEEL SET ON ITS CENTER

Illustration from Cecil Max Hilton, Equipment and Methods Used at Pulpwood Operations Upon the West Branch of the Penobscot River in Maine, 1935-40, University of Maine Thesis, 1940. Great Northern Paper Company Records, SpC MS 0210 Box 13, Folder 1, Raymond H. Fogler Library Special Collections Department, University of Maine, Orono, Maine

Newspaper columnist Lew Dietz wrote the captions shown below for these two boom jumper photos in the *National Fisherman*, May 1967, page 18-A. in an article entitled "Special Breed of Boats Worked Maine's Lakes."

GNP Co. *No. 41* at North Twin Lake
Great Northern Paper Company Records, SpC MS 0210, Raymond H. Fogler Library
Special Collections Department, University of Maine, Orono, Maine

"**TYPICAL SMALLER BOAT** on the GNP fleet was the *No. 41*, a 'boom jumper' type, many of which were built by Harkness. Stoutly constructed and then reinforced at crucial spots, these boats were at home among the pulp logs. They have been replaced by slightly smaller steel-hulled boats the profile of which closely resembles a Maine lobster boat." *National Fisherman* magazine, May 1967, p. 18-A

GNP Co. *No. 42* crossing boom at Seboomook Lake
Great Northern Paper Company Records, SpC MS 0210, Raymond H. Fogler Library
Special Collections Department, University of Maine, Orono, Maine

"**LIVING UP TO HER NAME**, a 'boom jumper' plows across a log boom on Seboomook Lake on the upper reaches of the West Branch of the Penobscot. This particular craft was not a Harkness design. Although smaller than the big West Branch class vessels, these little boats could haul sizeable booms of logs." *National Fisherman* magazine, May 1967, p. 18-A

To describe how boom jumpers crossed log booms, Dietz additionally wrote in that same *National Fisherman* article that "It took a special boat and a special man for the tricky job of jumping booms. Harkness designed the boats for the job and in due time there were men with the cowboy knack of handling them. When it was necessary to get inside a boom, the boat was headed directly at the barrier, throttle wide open. The trick was to hit the boom with the bow up so that the boat would rise smoothly on her forefoot and slide over the obstacle in one splashing leap." [22]

The Great Depression (1929-1939) brought GNP Co. boat building virtually to a halt. The West Branch waters remained an important and efficient way of transporting millions of cords of pulpwood logs to the GNP Co. mills, though, and "Admiral" Harkness's wooden-hulled boom jumpers worked well into the post-World War II years when they were gradually replaced by steel-hulled vessels of similar size and purpose.

Three GNP Co. steel-hulled boom jumpers, *No. 45*, *No. 46*, and *No. 49*, were built in the early 1950s at the General Foods Corp. shipyard at 79 Mechanic Street in Rockland, the former location of the Snow Marine Shipyard. General Foods Corporation was the parent company of General Seafoods Corp. which owned and operated the shipyard from 1949-1957. It built a sardine carrier at the facility, and records show that 11 "undocumented" steel tugs were also built there. [A documented vessel is one that is registered by the federal government through the U.S. Coast Guard, rather than titled and numbered by a state. Pleasure vessels of 5 net tons and over 26 feet in length and up may be documented, and commercial vessels 5 net tons and over must be documented. Since these boom jumpers were used on lakes and rivers, documentation was not necessary.] Three of these undocumented tugs went to GNP Co.; the other eight were sold to companies such as the Kennebec Log Driving Company in western Maine and the Brown Lumber Company in Berlin, New Hampshire, for log boom operations. [23]

Hull of GNP Co. *No 45* in 1951
Chesuncook Village Association website photo

GNP Co. *No. 45* was completed in April 1951 at the General Foods Corp. shipyard in Rockland, Maine, and transported to Chesuncook Lake. Its overall length was 35', beam 9', and depth of 4', and powered

48

by a 100 hp Lathrop diesel. It was transported the 200 miles to Chesuncook Lake by a tractor-trailer truck. Three other steel "log boom jumper" tugs, which followed in 1952-1954, were also built at the General Foods Corp. shipyard. All four were from the design plans of Camden naval architect Geerd Hendel.

GNP Co. No. 46 ready for launch at Rockland in 1952
Chesuncook Village Association website photo

GNP Co. *No. 46* was a steel "log boom jumper" tug launched at the General Foods Corp. shipyard in Rockland on April 18, 1952, to work at Chesuncook Lake. Overall length was 35', width 9', and depth of 4', and power was supplied by a 100 hp Lathrop diesel.

GNP Co. *No. 46*
Great Northern Paper Company Records, SpC MS 0210, Raymond H. Fogler Library
Special Collections Department, University of Maine, Orono, Maine

**Lobster boat type of steel-hulled "boom jumper" similar to GNP Co. *No. 46*,
except *No. 46* has a muffler and exhaust installed through the cabin.**
Great Northern Paper Company Records, SpC MS 0210, Raymond H. Fogler Library
Special Collections Department, University of Maine, Orono, Maine

**GNP Co. *No. 49*. This steel "log boom jumper" tug was launched at the General
Foods Corp. shipyard in Rockland in 1954 to work at Chesuncook Lake.
Overall length was 35', width 9', and depth of 4', and power was supplied by a
100 hp Lathrop diesel. Christened by Mrs. John Maines.**
Great Northern Paper Company Records, SpC MS 0210, Raymond H. Fogler
Library Special Collections Department, University of Maine, Orono, Maine

The descriptive information for vessels GNP Co. *No. 45, No. 46 and
No. 49*, which were built at the General Foods Corp. in Rockland, is
from Bertram Snow's book *The Main Beam*, published by the Rockland,
Maine, Historical Society. [24]

The task: keep the logs moving. Upper left — Driver uses pike pole to push log from shore. Center — Jam broken on West Branch. Right — A steel-hulled "boom jumper" smashes through an island of pulpwood on Chesuncook Lake. Photos by Glenn Richmond and Great Northern.

Steel-hulled "boom jumper" at work, breaking up a pulpwood "island" on Chesuncook Lake.
GNP Co. photo from Chesuncook Village Association website

bay Re

OF MAINE AND DEVOTED TO ALL ITS IN

Y HARBOR, MAINE, THURSDAY, APRIL 4, 1957

EAST BOOTHBAY YARD BUILDS STEEL TUGS FOR GREAT NORTHERN PAPER CO.

Hodgdon Bros., Goudy and Stevens, East Boothbay Shipyard, recently completed work and trial runs on one of three 35 foot steel tugs for the Great Northern Paper Company. The boat has a beam of 10' 2"; draft of 5' 2"; and is powered by a GM 6-71 Detroit Diesel, 4.5:1 reduction. The five blade propeller measures 48" and the boat has a capacity of 5,000 pound towing on bitt. There is a hydraulic throttle and clutch and Crowell hydraulic steering controls.

Similar models have been built at the yard previous to this group of three, all of them for the Great Northern.

These tugs are designed for rough service in handling logs. They are all welded steel construction and reinforced to withstand harsh treatment from boom log jumping and acres of pulpwood. They can handle tows of up to 5 acres of floating logs.

Hydraulic reverse and steering make for easy handling and maneuverability. In fact an important feature of this boat is the Crowell Steerer, as it enables the wheel to be left on course, unattended, while some other operation on the boat is carried out. This factor eliminates the need for an extra hand.

These boats have proved to be remarkably successful for the work they were designed to do, as the repeated orders for more have amply demonstrated.

This April 4, 1957, article in *The Boothbay Register* reports that GNP Co. had ordered three of these 35' steel boom jumpers which also had the capacity to handle log booms of up to 5 acres in size. They were fabricated by a Hodgdon Bros./ Goudy and Stevens boatyard partnership in East Boothbay. The article concludes with this sentence: "These boats have proved to be remarkably successful for the work they were designed to do, as the repeated orders for more have amply demonstrated."

Courtesy of *The Boothbay Register*, April 4, 1957, p. 1

This boom jumper, GNP Co. *No. 57* appears to be one of the vessels described above.
Moosehead Historical Society photo

Chapter 5

Other Great Northern Watercraft

Tethys II
The Northern, February 1927, pg. 6. Great Northern Paper Company Records, SpC
MS 0210 Box 21, Raymond H. Fogler Library Special Collections Department,
University of Maine, Orono, Maine

The original *Tethys* worked on Chesuncook Lake carrying woodsmen and supplies. [Tethys was the Greek goddess, wife of Oceanus, who became the mother of all the rivers of the earth.] This boat was built for Edwin Capen to replace a boat of the same name which was wrecked on Moosehead Lake when the vessel broke away from its mooring. The original *Tethys* was built as a pleasure boat on the Maine coast in the Sedgwick area but was then converted for use in the sardine trade. In 1888 Charles Capen purchased it and brought the boat to Moosehead Lake, where it was used for fishing parties. *Tethys* was 38 feet long, had a 10-foot beam, and was equipped with a steam engine.

The second *Tethys* was built in Brewer by Si Leach in 1895 for Charles Chapin and was brought to Moosehead Lake on the Bangor and Aroostook Railroad. This 52-foot boat carried passengers on Moosehead Lake for 20 years. In 1916 Chapin sold it to the GNP Co.

One of the first tasks given to O.A. Harkness, who began work for GNP in 1915, was to transport the *Tethys* from Moosehead Lake to the Canada Falls deadwater on the South Branch of the Penobscot River. In the autumn of 1916 it was put on greased ways at Seboomook (northwestern end of Moosehead Lake) and drawn by horses to Carry Pond thence to the West Branch of the Penobscot River, where it went to Pittston Farm under its own steam. At that point, it was taken out of the river again, put on greased ways over the ice and snow, and 16 horses hauled it to the Canada Falls deadwater. It sunk there on one occasion when it rammed a big spike on a pier. The *Tethys* worked there for about three years towing log booms, but it drew too much water and was replaced by a smaller boat.

In about 1919 it was returned to Moosehead Lake, this time hauled by two Lombard tractors. In 1923, it was taken out of Moosehead Lake at Lily Bay and pulled some 35 miles north by three Holt crawler tractors to Chesuncook Lake. In 1925, it broke loose from a mooring in a storm and was damaged again. At that point, it was sold by GNP Co. to Alec Gunn who rebuilt it and put it to private use. [25]

*The **Tethys** on the way to Canada Falls. O.A. Harkness, standing in the bow of the boat, is giving orders to drivers of the four teams of four horses each. The drivers were Robert Canders, J.P. Hayes, Hollis Rutledge and William Harrington.*

The Northern, February 1927, pg. 6, Great Northern Paper Company Records, SpC MS 0210 Box 21, Raymond H. Fogler Library Special Collections Department, University of Maine, Orono, Maine

On the trip from Lily Bay to Chesuncook Lake, the Tethys was drawn by three Holt tractors.

The Northern, February 1927, pg. 7, Great Northern Paper Company Records, SpC MS 0210 Box 21, Raymond H. Fogler Library Special Collections Department, University of Maine, Orono, Maine

Harkness prepared a design plan, dated December 27, 2017, for a 38' steam tow boat to be based on Penobscot Lake, the headwaters of the South Branch of the Penobscot near the Canadian border. There are no references to the vessel in any article in *The Northern*, or in any other source reviewed by the author, so the vessel may not have been built. (See plan on following page.)

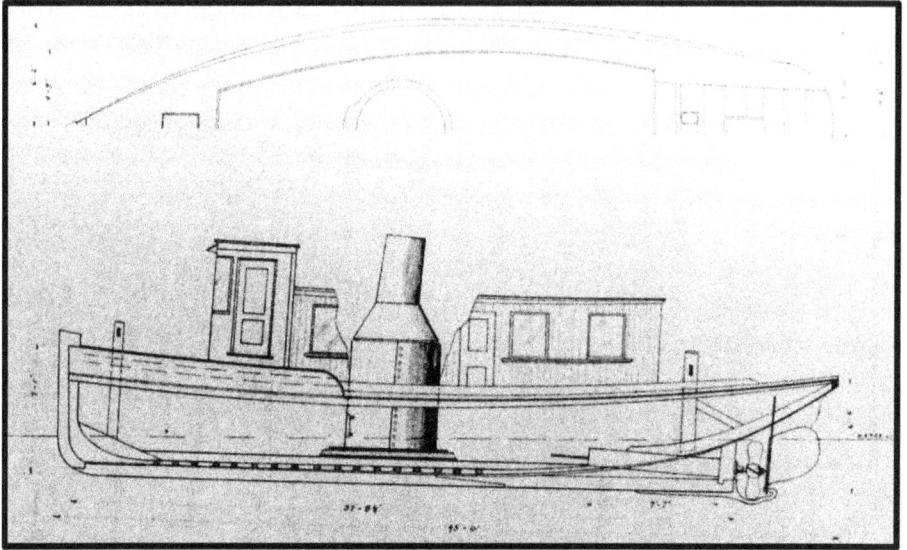

Great Northern Paper Company Records, SpC MS 0210 Box 13, Folder 1,
Raymond H. Fogler Library Special Collections Department,
University of Maine, Orono, Maine

GNP Co. *Ricochet*
The Northern, July 1925, photo, p. 9, Great Northern Paper Company Records,
SpC MS 0210, Box 21, Raymond H. Fogler Library Special Collections Department,
University of Maine, Orono, Maine

No detail has been found concerning the *Ricochet*'s use or work location, and GNP Co. workboats were typically given numerical names. Its design is not that of a workboat.

Charles Glaster, though, in his book *The West Brancher,* describes the GNP Co. vessel *Hope L.* as a pleasure craft purchased by GNP Co. to bring supplies daily to the five logging camps on or near Chesuncook Lake and transport workers among the camps. Glaster writes that the *Hope L.* caught fire on the bitterly cold night of November 21, 1920. It was transporting workers across 14 miles of Chesuncook Lake to the Cuxabexis Lake logging camp when the fire ignited. The boat sank near the shore, and the bow of the grounded boat was visible to rescuers. More than 30 workers were aboard; and although 16 of them were saved, the others tragically drowned.

Glaster mentions that a "new cabin boat, seaworthy and reliable" was provided as a replacement vessel in 1921, and perhaps that vessel was the *Ricochet.* [26]

Bateaux

As noted earlier, there were other, smaller, boats not of the boom jumper type in GNP Co.'s fleet, but the boom jumpers were the boats in which the 'Admiral' was chiefly interested. "For most purposes, even the bateau is giving way on the West Branch to the boom jumper which should bear the name of Harkness as its developer." [27]

Maine river-driving bateaux were flat-bottomed boats about 32' in overall length and made for use in rapids and whitewater as well as in shallow streams. They could be rowed in flat water but were usually maneuvered by using poles. They carried work crews to break up log jams and keep logs flowing on rivers during springtime log drives. They were fast and maneuverable, and of light-weight construction for carrying. The length of the bow and its flared shear kept it above any boiling whitewater, and the long rake of the stern allowed the boat to slide over obstructions. [28]

Bateau with crew at work on a log jam
Collection of the Patten Lumbermen's Museum

Bateaux with crews freeing log jams
Collection of the Patten Lumbermen's Museum

Chapter 6

The Heavy Tug *O.A. Harkness*

Vessel History, 1964-1972

In 1964, thirteen years after O.A. Harkness retired from the Great Northern Paper Company, the heavy-duty steel tug *O.A. Harkness* was designed by naval architect Geerd Hendel and built in 1964 for GNP Co. at Goudy and Stevens, Inc., in East Boothbay, Maine; the vessel is listed as hull #199 in their records. The cabin was built at Bath Iron Works (BIW) in Bath, Maine. The vessel was 70' long with a 20' beam and was powered by two 250 horsepower General Motors diesel engines. Two generators powered a 120-volt electrical system and heating system, and the vessel included a pilot house, dining area, crew's quarters, and head. Steel housings protected both propellers against damage from floating or submerged logs. When completed, the vessel cost approximately $125,000.

O.A. *Harkness* cabin being loaded at Bath Iron Works, Bath, Maine
Great Northern Paper Company Records, SpC MS 0210, Raymond H. Fogler Library
Special Collections Department, University of Maine, Orono, Maine

This article about the launching of the *O.A. Harkness* and honoring Harkness himself appeared in the "Great Northern Salaried Employees Newsletter" of June, 1964. "The new towboat on the lower lakes will be officially launched and named the *O.A Harkness* on June 6. O.A. Harkness was a valued employee of the Company from 1917 [*sic*] to his retirement in November 1950. Though he lacked much in formal education, he was a mechanical genius of the highest order. Towboat *West Branch No. 3*, which the *O.A. Harkness* replaces, was built under Mr. Harkness' supervision and launched in May, 1948. A list of his inventions and improvements on equipment would take pages to list. Above all, he was a man who inspired others and who earned the respect of all who came into contact with him. The *O.A. Harkness* will be formally christened by his widow."

O.A. Harkness being christened by Mrs. Alice Harkness, June 6, 1964
Great Northern Paper Company Records, SpC MS 0210, Raymond H. Fogler Library Special Collections Department, University of Maine, Orono, Maine

The "Great Northern Monthly Salaried Employee Newsletter" further reported in its July 1964 issue that on Saturday, June 6, Mrs. Alice Harkness, widow of the late O.A. Harkness, cracked a champagne bottle on its bow, wishing Godspeed to the boat *O.A. Harkness* and its crew. The ceremony took place at Norcross on the "lower lakes" near Millinocket. About 300 people were in attendance and enjoyed a typical woodlands dinner. Many of those present were named "honorary admirals" in GNP Co.'s landlocked navy.

O.A. Harkness, 1964
Great Northern Paper Company Records, SpC MS 0210, Raymond H. Fogler Library
Special Collections Department, University of Maine, Orono, Maine

A crew of six operated the boat, originally on a 24 hour per day basis, navigating by compass, searchlight, and skyline landmarks, but towing was later limited to daylight hours.

O.A. Harkness Sold to the State of Maine

Great Northern Paper Company estimated that the *O.A. Harkness* had handled about 120,000 cords of pulpwood annually in the "lower lakes," navigating those restricted waters with an average of 2,500 to 3,000 cords in each tow. Figuring 60 to 65 logs per cord, the Harkness had towed enough pulpwood, if placed end to end, to reach about 46,000 miles, equal to a chain of logs circling the earth twice. In order to tow the pulpwood, a double loop or "bag" of boom logs 28-30 feet in length, tied together with heavy chains, was created. As noted in Chapter 2, towing across the open lakes was relatively simple, although the tug operated at one mile per hour. A higher speed would cause the pulpwood logs to jump over the back edge of the boom. Maneuvering booms through narrow channels could be tricky and limited the number of logs that could be towed at one time. Wind storms also could cause an entire tow to be abandoned until the weather moderated.

The *O.A. Harkness* became available for purchase when GNP Co. phased out its river log drives in 1971, due to the enactment of environmental protection laws. GNP Co.'s river log drives began in 1899, but after 1971, all pulpwood was brought to the paper mills by truck.

The ruggedly built craft, judged to be well suited to the operational needs of the Maine Department of Sea and Shore Fisheries, was purchased by the state in 1972. At the time of purchase, the *O.A. Harkness* was located on North Twin Lake near Millinocket, having been used to tow log booms on the upper Penobscot River across Ambajejus Lake, Pemadumcook Lake, North and South Twin Lakes, and Elbow Lake. This was a distance of about 14 miles, and between 2,500 and 3,000 cords were brought to the Great Northern mill in each tow. [29]

After it was sold, the cabin was separated from the hull, and the two sections were transported by road to Bangor. It was reassembled and launched into the Penobscot River to make its trip to Boothbay Harbor where it was based at the Sea and Shore Fisheries facility.

Harold MacQuinn, Inc. acquires the *O.A. Harkness*

The *Harkness* was later sold by the state to Harold MacQuinn, Inc., a construction firm based in Bar Harbor, and used to tow construction company barges around the Maine coast. It was apparently engaged in such a tow when it began taking on water on January 16, 1992, just two miles south of Matinicus Island.

The leak seemed to be coming from a snapped rudder post. The *O.A. Harkness* crew of three was involved in a harrowing sea rescue, which occurred after dark, in seas with eight-foot swells, and a wind chill temperature of -54 degrees.

While the boat was gradually sinking, the crew was able to transmit their location via radio to lobstermen who were listening to the marine radio on Matinicus Island, and Vance Bunker and two friends set out in Vance's lobster boat to attempt a rescue. A Coast Guard boat was also dispatched from the Rockland Coast Guard Station. The *Harkness* sank. When the three men wearing immersion survival suits abandoned the *Harkness*, one of the men grabbed a flashlight. The three were together in the water clinging to a bit of wreckage and the flashlight had gotten turned on, shining upward through the sea smoke. At that moment, the flashlight beam was quickly seen shining by those on the lobster boat and the three *Harkness* crew, Captain Rudi Musetti, mate Arthur Stevens, and crewman Duane Cleaves, were rescued and taken to Matinicus Island. [30]

The construction of the *William Hilton* for GNP Co. in 1961 preceded that of the *O.A. Harkness* by three years. Both vessels were built from the same Geerd Hendel plans and both hulls were constructed at the Goudy and Stevens Boatyard in East Boothbay, Maine. Both vessels were designed to tow booms of 8,000 cords of pulpwood.

The *Hilton* (Goudy & Stevens, Inc. hull #192) was based on Chesuncook Lake, replacing the *West Branch No. 2*. It was named for GNP Co. Vice President of Engineering and Woodland Management, William Hilton, who retired in 1959 after 47 years with the company. Hilton began his working career as the first watchman at the Moose Mountain fire tower near Greenville, which was the first such tower established in the state.

The Heavy Tug *O.A. Harkness*

***William Hilton* being trucked to Chesuncook Lake in 1961**
Great Northern Paper Company Records, SpC MS 0210, Raymond H. Fogler Library
Special Collections Department, University of Maine, Orono, Maine

William Hilton
Great Northern Paper Company Records, SpC MS 0210, Raymond H. Fogler Library
Special Collections Department, University of Maine, Orono, Maine

Log boom tow cable and winch aboard the *William Hilton*
Robert Laverty Collection, Northeast Historic Films

For Hendel-designed winches on similar tow boats, the engine was equipped with a front-end power takeoff geared through a three-speed transmission to operate a warping winch down in the engine room. The log boom tow winch drum had a capacity of 4000' of 5/8" steel cable.

Chapter 7

The Tramway – Eagle Lake to Chamberlain Lake

In the 1830s, Amos Roberts and the Strickland Brothers bought Township 6, Range 11, a piece of land that contains the drainages into two major watersheds. Waters of Webster Lake, a natural headwater of the East Branch of the Penobscot River, flow southerly to Bangor. The waters of Telos Lake, which in its natural state is a headwater of the Allagash River, go northeastward and merge with the St. John River which flows into New Brunswick, Canada (see map on page 5). In order to get logs harvested from the area around Telos and Chamberlain Lakes down the East Branch where American interests could profit, the timber owners needed to devise a way to move the logs from Telos over to Webster Lake.

In 1838, they engaged Shepard Boody to devise a way to get Chamberlain Lake to flow against its natural current into Telos Lake and thence into the East Branch basin. Boody proposed raising the levels of Chamberlain Lake and Telos Lake by about 11' with dams and digging a canal across part of the area between Telos and Webster Lakes. Beyond this area, at the foot of the Telos Dam, was a ravine that dropped approximately 47' into Webster Lake. By fall 1841, the two dams were in place and a canal 10' to 15' wide and 1' to 6' deep connected Telos to Chamberlain Lake. It was thereafter known as the Telos Cut.

The two dams enabled United States timber owners to control the flow into the canal Cut and thus collect a toll per 1,000 board feet of lumber from any landowner who wanted to drive their logs to Penobscot River mills and markets. However, it also made them dependent on the ability of the Chamberlain Lake dam owners to retain enough water in Telos Lake to force logs to move southward to the Penobscot. This led to a series of controversies over toll charges between owners of the Telos Dam and Cut and the owners of the Chamberlain Lake Dam, called the "Telos War," which ultimately had to be settled by an act of the state legislature. [31]

Note: These two dams are now maintained and managed by Maine Department of Agriculture, Conservation and Forestry, Bureau of Parks and Public Lands, as part of the Allagash Wilderness Waterway.

In the 1850s, the Chamberlain Dam was rebuilt to include a series of locks that were used to float groups of logs from Eagle Lake to Chamberlain Lake. To this day, the dam continues to be known as the Lock Dam. The lock process was slow, however, and the idea of constructing a tramway to cross the 3,000' land bridge between the two lakes was developed.

At top, Webster Lake, geographic headwaters of the East Branch of the Penobscot

At center, Telos dam and Telos cut, which cause Chamberlain and Telos waters to flow southerly into the East Branch instead of northerly into the Allagash watershed.

At bottom, Telos Lake which connected to Chamberlain Lake

Collection of the Patten Lumbermen's Museum

In 1901 pulpwood harvesting was planned for the Eagle Lake area (north of Chamberlain Lake) in the north-flowing Allagash and St. John River watersheds running into Canada. The obstacle to bringing the logs from this Eastern Manufacturing Company timberland into the East Branch and main stem of the Penobscot to send them to the Company's saw and paper mills in Brewer was a 3,000-foot wide isthmus separating the waters of Eagle Lake from those of Chamberlain Lake.

To overcome that obstacle, a tramway was proposed. The tramway was designed and constructed by engineer Fred T. Dow, based on the creative idea of lumber barons and PLDC owners H.W. Marsh and F.W. Ayer (Ayer also owned the Eastern Manufacturing Company paper mill in Brewer). The tramway was composed of a small-gauge elevated railroad track on which ran 600 steel-wheeled "trucks" which carried logs across the 3,000-foot passage, attached to and pulled by a 6,000-foot continuous loop cable.

Most of the tramway machinery was boated across Moosehead Lake in the fall of 1901 to Northeast Carry and then during the winter placed on sleds and hauled by horses overland 42 miles northerly to Eagle Lake. It was a terribly difficult, exhausting, task through very remote timberlands. Although the intention was to move the 14-ton cable as a continuous strand on two drums using horse teams and skids, it eventually did have to be cut into two sections to lighten the load to get it to its destination. [32]

The tramway was a complicated system of tracks, cables, and engines to pull pulpwood logs from Eagle Lake across 3,000 feet of land surface and slide them into Chamberlain Lake for towing. It ran at the rate of about three miles an hour, or about 250' per minute. As the logs dropped off the elevated portion of the track at the Chamberlain Lake end, each empty "truck" looped underneath to a ground level track and was pulled by the cable back to Eagle Lake to be reloaded.

Harkness had been hired by the Penobscot Log Driving Company in November 1901 and was assigned in early 1903 to build the side-wheeler *George A. Dugan* at Chamberlain Lake. In April of 1903, F.W. Ayer, in his role as part owner of the PLDC, wrote to Harkness asking him to look after finishing up the company's Eagle Lake to Chamberlain Lake tramway and putting it into full operating condition. The tramway had significant startup difficulties, and Harkness agreed to undertake the challenging task. [33]

For his work to successfully address both the initial mechanical and operational problems of the tramway, O.A. Harkness quickly developed the reputation of being a "mechanical genius." [34]

Author and historian David C. Smith writes the following about the start-up of the tramway operation: "On the first run 4,800 bolts and clamps that had not been threaded deeply enough stripped their threads and had to be regrooved. Later during that run, Herbert Marsh who was supervising the loading of the logs became insistent that the men put a log in each clamp. The weight was too much, and the cable kept stopping. O.A. Harkness the ostensible straw boss came from the Chamberlain Lake side to see what was causing the delay. Marsh asked him as he approached if there was anything he could do. According to the testimony of the people who saw the exchange, Harkness answered, 'Yes, go back to Bangor and let us handle the damn logs the way they ought to be handled.' Marsh in any event went back to Bangor and the logs began to move. For the next six years, the tramway moved 500,000 feet every working day, or about fifteen or sixteen million a year. Harkness later estimated that 100,000,000 feet went over the tramway and down to Bangor mills, wood which would have otherwise gone to Fredericton (New Brunswick, Canada) by the St. John River. Yankee ingenuity was able here to extend the lumbering era on somewhat longer even after the press of economics had forced the coming of the pulp paper revolution elsewhere." [35]

In the November 1927 issue of *The Northern*, Harkness described the tramway this way: "The steel cable, 1-1/2" in diameter, was 6,000 ft. long and fastened so that it was endless and reached from Eagle Lake to Chamberlain Lake. At intervals of 10 ft. the trucks were clamped on. These trucks consisted of a steel saddle on which the log rested, and two 11" wheels which ran on steel rails apart. There were two tracks, one above the other. The loaded one went on the top track and the empty one returned on the lower track. Halfway between the trucks there was a steel clamp. Both the clamp and the truck fitted into the sprocket wheel, which was 9 feet in diameter, situated at the Chamberlain end of the tramway. This sprocket wheel made nine revolutions per minute which made the log travel at the rate of 250 feet per minute. The sprocket wheel was geared to a Westinghouse Compound Engine designed especially for electric light plants. The cylinders were 12" and 24" with a 14" stroke. The engine made 255 revolutions per minute with 100 pounds of steam. Wood was fuel for the two boilers furnishing steam for the engine. It took a lot of power to start the machinery moving but it rolled easily once it was in motion.

Note: A compound steam engine unit is a type of steam engine where steam is expanded in two or more stages. Typically, in a compound engine the steam is first expanded in a high-pressure cylinder. Then having given up some heat and losing pressure, it exhausts directly into one or more larger-volume low-pressure cylinders, to extract further energy from the steam.

The tramway was used for six seasons. It was under my care for the entire time. We averaged about 500,000 feet, board measure, for each operating day beginning at 4 A.M. and ending about 8 P.M. This tramway was always an excellent piece of machinery and very efficient in doing the work for which it was designed. The total amount of lumber taken over the tramway was about one hundred million feet." [36]

The tramway ran right into the water of Eagle Lake and turned a submerged wheel. Here men with caulked boots and pick-poles pushed and shoved the logs into place so the tramway trucks could pick up the logs as they came out of the water. On the Chamberlain end a spur wheel picked up the logs from the tramway trucks and pushed them down a slip into the lake. [37]

On Eagle Lake each spring and summer, the *H.W. Marsh* towed boomed logs, harvested during winters and hauled by horses and sleds across the snow to the shore of the lake, to the tramway. At the Chamberlain Lake end of the tramway, the logs were then re-gathered into log booms and towed by the *Charles A. Dugan* across Chamberlain Lake for driving down the East Branch of the Penobscot. (See Chapter 3 for details of these two vessels.)

Note: Because the East Branch enters the main stem of the Penobscot below Millinocket, the logs were pulled from the East Branch at Grindstone and hauled seven miles to the West Branch, first by horses, in the 1910s by Lombard log haulers and 1920s by Holt crawler tractors. [38]

74

View of empty tramway "trucks" at Eagle Lake, with guide logs placed along both sides
Collection of Patten Lumbermen's Museum

BOB MOORE REPORTS TO MR. HARKNESS

O.A. Harkness at tramway
The Northern, April 1927, pg. 9, Great Northern Paper Company Records, SpC MS
0210 Box 21, Raymond H. Fogler Library Special Collections Department,
University of Maine, Orono, Maine

One of the 600 tramway "trucks" attached to the tramway cable
Photo by the author 2017

**A reconstructed tramway section at the Chamberlain Lake end of the tramway
showing the two-level wood cribwork with a log placed across two trucks**
Photo by the author 2017

A Westinghouse compound engine and two steam boilers provided the power to pull the 3,000' of 1 1/2" cable across the land barrier between the two lakes. Logs were floated from the waters of Eagle Lake onto the top row of the toothed "trucks" and pulled by the cable along the upper level of the wooden structure for the 3,000' distance to Chamberlain Lake. At the Chamberlain Lake end, the logs were pushed down a sluice into Chamberlain Lake and the logs floated free. The empty "trucks" were pulled back by the continuous-loop cable along the lower level of the structure for the 3,000' return trip to Eagle Lake, with the whole process then repeated continuously. The sprocket wheel, 9' in diameter, remains sited at the Chamberlain end of the tramway. It rotated nine times per minute which pulled the logs carried on the tramway across the isthmus at the rate of 250' per minute. Halfway between each of the 600 trucks, there was a steel clamp attached to the cable. Both this clamp and the truck fitted into this sprocket wheel to enable the cable/trucks to be pulled in a continuous loop back to Eagle Lake. The sprocket wheel was geared to a Westinghouse Compound

Engine designed especially for electric light plants. Leather belts, two feet wide, 60 feet long, and weighing over 200 pounds, ran from the engine to the rollers on the right-hand side of the wheel (see above) to move the gears that drove the tramway cable. [39]

9' sprocket wheel situated at the Chamberlain end of the tramway
Photo by the author 2017

Power house at the Chamberlain Lake end of the tramway
Maine Department of Conservation

The tramway was rendered obsolete by the invention of the Lombard Steam Log Hauler (see Chapter 8), and ceased operation in 1907. The Patten Lumbermen's Museum in Patten, Maine, has in its collection a set of two of the tramway's tracks and clamps mounted on a wooden structure as it would have been during actual operation.

In 1927 the Eagle Lake end of the tramway site was re-adapted for use as the eastern terminus of the Eagle Lake and West Branch Railroad, a rail transport system for hauling timber southward to the West Branch (see Chapter 10).

The site and remnants of the Tramway (1901-1907), and the Eagle Lake and West Branch Railroad (1927-1933) terminus on Eagle Lake, which re-used the Eagle Lake end of the Tramway site, were listed on the National Register of Historic Places as the Tramway Historic District in 1979. The District is now part of the 145,000 acre (land and water) Allagash Wilderness Waterway, a Maine state park purchased with a combination of state and federal funds in 1965.

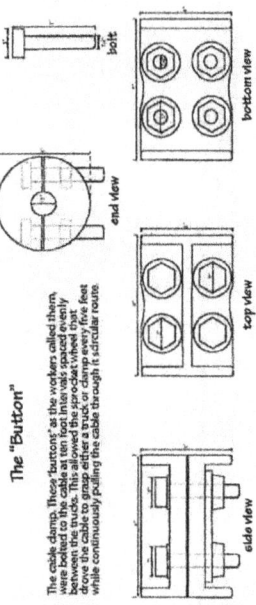

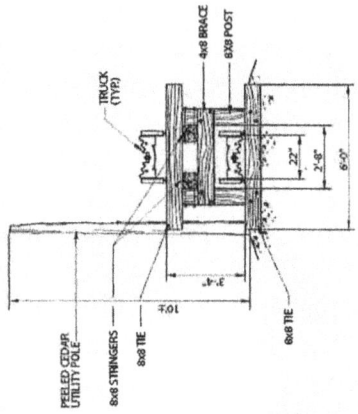

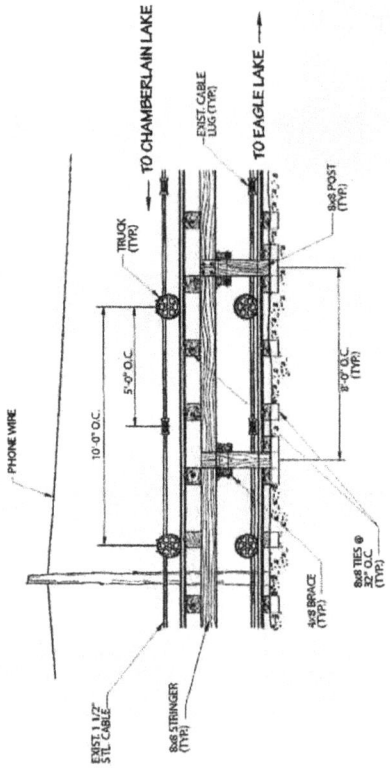

The "Button"

This cable clamp. These "buttons" as the workers called them, were bolted to the cable at ten foot intervals spaced evenly between the trucks. This allowed the sprocket wheel that drove the cable to grasp either a truck or clamp every five feet while continuously pulling the cable through it's circular route.

EAGLE LAKE TRAMWAY (1902-1907)

TRAMWAY CRIB STRUCTURE

SECTION

TYPICAL ELEVATION

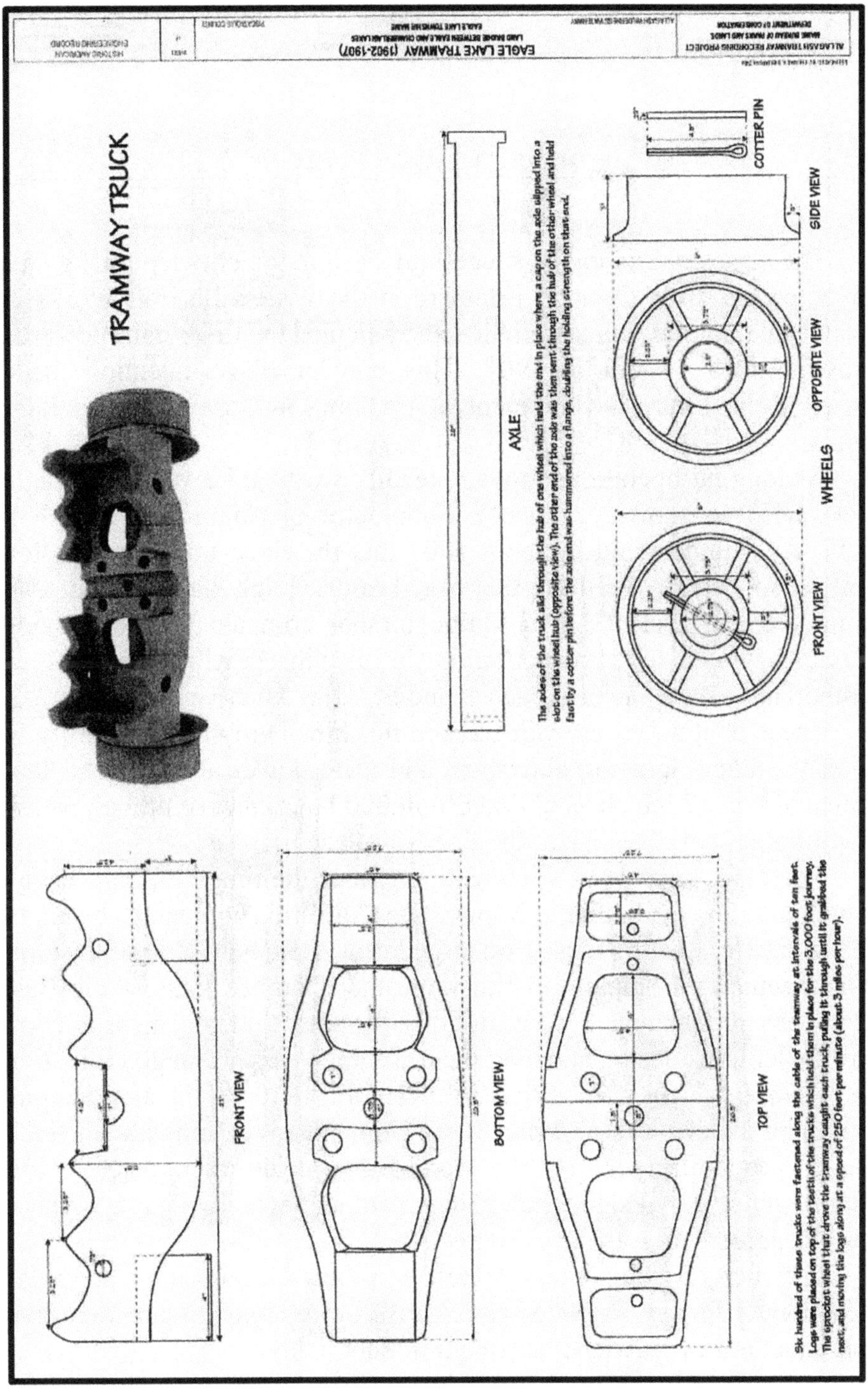
EAGLE LAKE TRAMWAY (1902-1907)
TRAMWAY TRUCK
AXLE
COTTER PIN
SIDE VIEW
OPPOSITE VIEW
WHEELS
FRONT VIEW
FRONT VIEW
BOTTOM VIEW
TOP VIEW

Chapter 8

Lombard Log Haulers

The first commercially successful design for crawler tracks was invented by Alvin Orlando Lombard at the Waterville Iron Works in Waterville, Maine, who assembled and patented the first steam-powered crawler track system in 1901. This crawler track invention made possible the future development of tractors, bulldozers, and military tanks.

As logging operations moved steadily westward away from Eagle Lake and the tramway, Eastern Manufacturing Company's President F.W. Ayer understood in about 1907 that the fixed tramway location was obsolete. He realized that the Lombard log hauler, first sold commercially in 1903 to a Maine lumber company, could provide flexible winter log transportation in the inaccessible portions of timberlands in the upper Allagash and St. John River watersheds.

These continuous crawler-tracked machines provided flexibility in their woodland location placement and were powerful enough to haul trains of 8 to 12 log sleds of 24,000-40,000 board feet of timber, which contributed to the demise of the fixed-in-place tramway. [40]

Note: As an aside, it's interesting that at the time Lombard began developing his log hauler in Maine, two California companies began to make crawler tractors for use on large farms. Benjamin Holt of the Holt Manufacturing Company in Stockton and Clarence Best of the Best Company of Oakland, sued each other for patent infringement relating to crawler track design and over the next nine years prolonged litigation took place which was settled in 1919. In 1910 Alvin and Samuel Lombard traveled to California to see Holt and advise him that his track design was infringing on the Lombard crawler track patent. Holt reportedly acknowledged the infringement and agreed to pay a royalty, but the agreement was never signed. [41]

The merged companies of Holt and Best became the Caterpillar Traction Company in 1925, which, of course, remains an important manufacturer of crawler-tracked equipment today.

Records show that on December 25, 1907, the Eastern Manufacturing Company purchased two steam log haulers directly from the factory in Waterville, and later purchased two additional machines for their Allagash River operation.[42]

O.A. Harkness was interviewed by *The Northern* for the December 1927 issue, in an article entitled "Log Haulers 20 Years Ago." Twenty years prior, he had been the Master Mechanic for the Operations of the Eastern Manufacturing Company, starting in November 1901, and was assigned to the Penobscot Log Driving Company in 1903 to activate and operate the tramway which remained in use for six years. He reported that in 1907 it was decided to try out three Lombard steam log haulers, and a 12-mile long log hauler road was constructed in the upper Allagash watershed (in Township 9 Range 17) near the head of Chamberlain Lake. It was found to be cheaper to the haul trains of timber to the shore of Chamberlain Lake with the Lombards than to land it in Eagle Lake and tow it over to the tramway.

With tracks at the rear and skis for steering at the front, the Lombards were a winter operations vehicle which could efficiently and flexibly haul logs on roadways iced by huge sprinkler sleds 32' long and 12' wide, each holding 13 thousand gallons of water. The icing operation usually took place at night. There were three-inch holes in the rear of the tanks with plugs that could be removed to regulate the water falling onto the runner tracks.

The early steam log haulers could not stand as much rough usage as those used in 1927. Many parts were made of cast iron which would break very easily in really cold weather. The parts were later made from manganese steel which was not easily broken, so the log haulers became much more durable. [43]

About eight or nine Lombard log haulers were used by GNP Co. in the wood harvesting option over a period of six years.

Lombard Log Hauler with tree-length logs
Collection of Patten Lumbermen's Museum

Lombard Log Hauler with pulpwood logs
Great Northern Paper Company Records, SpC MS 0210, Raymond H. Fogler Library
Special Collections Department, University of Maine, Orono, Maine

Lombard log hauler
Collection of Patten Lumbermen's Museum, Patten, Maine

The purpose of the hose draped across the boiler was to draw steam boiler water from rivers and streams at Lombard's regular intervals. Water towers were later constructed to decrease the waiting times and thus increase their log hauling efficiency.

Steam Lombard ready to take on steam boiler water
Chesuncook Village Association website photo

Lombard Snow Plow

Harkness had constructed a homemade V-shaped snow plow out of logs for a Lombard and told this light-hearted story of a ride on it. "John Kelley and I started one afternoon to go to the landing on the snow plow. He was standing on one side and I on the other. The engineer was running her fast. The snow had been piled up on each side by the snow plow so that the banks were four or five feet high. The snow plow struck a stump, John Kelley went into the air, turned over once, cleared the high banks of snow and was buried in the soft snow beyond his side of the plow, while I did the same on my side! When we dug out and crawled back on, John was very much provoked. He called the engineer a variety of names! Now, every time we meet, some reference is made to our ride on the snow plow." [44]

In the end the Lombard log hauler lost the race with technology. The development of heavier and more durable trucks with pneumatic tires enabled great flexibility in wood harvesting operations. Additionally, the development of crawler-tracked bulldozers and other earth moving equipment made the construction of all-weather roads practical. No longer was the movement of timber restricted to land and ice in winter, and lakes and rivers in summer. [45]

1940s era logging trucks
Chesuncook Village Association website photo

Chapter 9

Crawler Tractors Replace Lombard Log Haulers

The U.S. history of tracked crawler tractors apparently began in California farm country, where some early experimental models were built. Alvin O. Lombard in Waterville, Maine, however, was the first commercially successful manufacturer of crawler tracked equipment with his patented crawler track propelling the Lombard Log Hauler. In California, Benjamin Holt had established the Holt Manufacturing Co. and began designing his own crawler tracks based on Lombard's, selling its first steam-powered crawler tractor about 1907. [46]

The Holt Manufacturing Company, and the California-based Best Manufacturing Company were direct competitors. In 1905, they resolved a patent infringement lawsuit with each other and the two companies merged in about 1908. Holt registered "Caterpillar" as a trademark in 1911, and the merged company became the Caterpillar Tractor Company in 1925 (see Chapter 8).

In 1917 GNP Co. started experimenting with crawler tractors for toting wood, but the early models of the Holt, Best, and Lombard crawler tractors proved unreliable in snow and below zero weather. [47]

"It was probably in concert with O.A. Harkness in 1921, then an executive of Great Northern, that Lombard resolved to produce a full track machine, called at the time a 'tank type' tractor and named at the shop the A-O-L, either for the initials of Alvin Orlando Lombard or for 'on-all-lags,' as Ralph Bickford, [an] old time employee believed." [48]

Beginning in about 1922, tank-type tractors with crawler tracks gradually came to replace the Lombard log haulers in Maine timber harvesting. With their flexibility in traversing uneven terrain to haul logs and operating with diesel or gasoline fuel, they enabled wood harvesting operations to stretch ever farther from lakes and train tracks. GNP Co. purchased Holt gasoline powered crawler tractors as early as 1923, as demonstrated when it used three of them in tandem to tow the *Tethys* 35 miles from Moosehead Lake to Chesuncook Lake (see Chapter 3).

These Holt Model 010 crawler tractors were built between 1923 and 1925, and the notation on this Maine Forest Service photo identifies it as "GREAT NORTHERN TRACTOR - RIPOGENUS, ME"
www.mainememory.net

Harkness wrote a brief article in *The Northern* issue of June 1922 describing a new design of "tank type" tractor by The Lombard Traction Engine Co. "… this machine will fulfill the requirements where the old type of machine cannot be used. This machine is known as the A-O-L type and has given some very good demonstrations. The design varies considerably from the tank type machine now in use. The engine is a duplex, four cylinders on each side. Two crank shafts are in one case. There are two clutches and two transmissions. The tracks are driven independently from each other. The steering is accomplished by the engine. It works out very satisfactorily. The tractor has been sent to Greenville Jct., to the Great Northern Paper Co., and is now in their shop waiting for a set of summer tracks. The Great Northern are to give this tractor a thorough testing and if it works out to their satisfaction they will no doubt adopt it as their standard machine." [49]

Lombard A-O-L tractor in 1922
The Northern, June 1922, pg. 13, Great Northern Paper Company Records, SpC MS
0210 Box 21, Raymond H. Fogler Library Special Collections Department,
University of Maine, Orono, Maine

Conquering the deep snow-drifts. The cab is temporary

New Lombard "New Twin" tractor in 1927
The Northern, April 1927, pg. 7, Great Northern Paper Company Records, SpC MS
0210 Box 21, Raymond H. Fogler Library Special Collections Department,
University of Maine, Orono, Maine

Significant engineering modifications to the Lombard tank type of tractor were needed to suit GNP Co.'s wood harvesting and other operational requirements.

Five years later, in *The Northern* edition of April 1927 (p.7), Greenville Machine Shop Superintendent Frederick Schenck wrote as follows:

" 'She has plenty of power,' declared Mr. O.A. Harkness, as he stood watching the New Twin Tractor bucking her way through the hardest of Greenville's snowdrifts. When the New Twin Tractor hauled two sleds of coal to the top of Blair's Hill on Feb 1, 1927, a long-cherished dream was realized. Back in 1921, our company ordered a special machine that would meet the requirements of all-round use. There was no machine of the tank type on the market, with the power and speed desired.

A tractor, which we here will call the Old Twin, was designed by and built by the Lombard Traction Engine of Waterville, Maine, for the Great Northern Paper Company. Two four-cylinder Stearns motors with a single crank case were the power units. One engine rotated clockwise

92

and the other counter-clockwise. The transmissions and rear ends were mounted in one big case. It had two clutches and two transmissions; the tracks were driven independently of each other. Steering was accomplished by suitable means for speeding up one engine and simultaneously retarding the other, thus causing changes in directions without losing power in the lag belts or treads.

This tractor was tried out during the summer of 1922 and proved to be superior to any machine available, but it had weak points. Mr. O.A. Harkness was satisfied that the principle was right, but that changes would be necessary to make the machine come up to his expectations. The tractor was torn down and the engine transmission case and other parts discarded.

Plans for the New Twin were begun in 1923. Interruptions were numerous and it was November of 1924 before real work on the machine was under way. In January 1925, H.A. Woodruff of the Union Iron Works, Bangor, came as draftsman. The steel casings, designed by Mr. Woodruff under Mr. Harkness' direction, were made and shipped to the Lombard Traction Engine Co. for machining. The assembling began at the Greenville shop in Sept. 1925. Delays were common 'while waiting for the castings,' until the most optimistic workmen were discouraged. Success was at last attained, and it was a proud day for the men of the shop when the New Twin made a trip around the shop yard. On Feb 1, 1927, it took two sleds of coal to the top of Blair's Hill [in Greenville]. The following day it took four loads of coal to Lily Bay [east side of Moosehead Lake]. Since then it has been used constantly for general toting.

The New Twin has two Wisconsin motors mounted on special hangers at 8 degrees. The motors have a 5 3/4" bore with a 7" stroke, each engine developing 63 H.P. at 800 R.P.M. Each side has an engine, brake and rear end gears mounted independently. This has resulted in remarkable flexibility in steering. A wheel, not unlike the steering wheel of an automobile controls the throttles of both engines. [O.A. Harkness's U.S. Patent #1,805,141 approved on May 31, 1931, for a steering wheel mechanism for a two-engine tractor, formally recognized and protected this remarkable steering system design. See Chapter 12].

When the wheel is turned in either direction, it throttles down the engine on that side, changing the direction of the tractor to that angle.

93

The tractor has a speed of eight miles per hour in high gear and 3 miles in low ....

Mr. Harkness insisted on exceptionally strong and rugged gears, cases and frame. He also secured large clearance, even balance, speed and the ability to travel through rough country and deep snow.

During the past month, the tractor has demonstrated its power, speed and flexibility. It has the speed and steering of the long base auto type tractor plus the ability of the tank type tractor to turn in short space and go over rough ground. Such a combination has not heretofore been effectuated. Mr. E.W. Englebright, President of the Lombard Tractor and Truck Corporation, and their consulting engineer, Mr. S.L.G. Knox, recently witnessed an exhibition of what the tractor can do. They expressed themselves as being well pleased with its performance and consider it an ideal tractor for woods use." [50]

Terence Harper in a February 2017 article available online wrote that the Lombard rebuilt "New Twin" A-O-L tractor met all of GNP Co.'s requirements, used Harkness's patented steering system, and was put to immediate use. "The last record we have of the New Twin – the only one of its kind ever built by Lombard, was in 1928. By then it had been equipped with a snow plow designed by Samuel Lombard and was praised for its service in opening up the road from Greenville to the Grant Farm. The A-O-L's unique duplex drive would not come to light again until the introduction of the massive twin engine Euclid TC-12 bulldozer in 1955." [51]

During the mid-1920s Lombard went through bankruptcy and its marketing shifted from timber harvesting to the construction and municipal markets.

Holt tractor hauling log sleds in late 1940s or early 1950s.
Great Northern Paper Company Records, SpC MS 0210, Raymond H. Fogler Library
Special Collections Department, University of Maine, Orono, Maine

Note the parallel return track which allowed empty sled trains to return without disrupting the flow of loaded sleds.

Property of Great Northern Paper, Inc.

GNP Co. undated photo of crawler tractors in a logging operation
Great Northern Paper Company Records, SpC MS 0210, Raymond H. Fogler Library
Special Collections Department, University of Maine, Orono, Maine

Chapter 10

The Eagle Lake and West Branch Railroad

The Madawaska Company employed as many as 3,000 men, many of them French-speaking Canadians, harvesting wood in the north woods of Maine. Its Canadian owner, Edouard "King" Lacroix, was known for his honesty and fairness. He paid his men well, furnished the best equipment, provided clean and comfortable living quarters, and fed his lumbermen as they liked to be fed. In 1925 Lacroix contracted with GNP Co. to deliver 125,000 cords of wood from the Eagle Lake-Churchill Lake region to the waters of the West Branch of the Penobscot to reach the GNP Co. mill in Millinocket. The area to be harvested was extremely remote, miles from any road or railroad.[52]

The initial harvesting plan had been to drive the wood down the Allagash to the Bangor and Aroostook (B & A) Railroad in Van Buren in northern Aroostook County. A dispute over B & A freight rates, however, led Lacroix to propose transporting GNP Co.'s wood into Chamberlain Lake (using part of the former tramway site) and sending the wood down the East Branch to Millinocket. [53]

> Note: The paper mill constructed at Millinocket on the West Branch in 1899 was served by the B & A Railroad via a spur also built that same year, and was used for shipping finished paper products nationally and internationally. The railroad also served the new, large GNP Co. mill built seven miles downstream at East Millinocket in 1906.

In early 1925, King Lacroix, F.A. Gilbert [GNP Co. Vice President Fred A. Gilbert], O.A. Harkness, and A.V. MacNeil made a canoe trip down the East Branch of the Penobscot to take one last look at that branch of the river before committing to the West Branch. This group of "able, practical men" found that preparing the East Branch for this scale of log drives was cost-prohibitive due to several large drops in the river which would have required several large dams to be built at

97

strategic points. The West Branch had long been used for log drives and had no such obstacles. [54]

With the combination of their long experience with and understanding of operating costs, marketing, pulpwood transportation systems, and the geography of the north woods, Gilbert, Lacroix, and Harkness conceived an even more visionary and ambitious scheme: they planned a logging railroad through the wilderness from Eagle Lake to Umbazooksus on the West Branch. Strategically for GNP Co., this plan kept the pulpwood transportation routes and system within its own control, and offered lower transport costs than any other option then available. It updated the Eagle Lake (Allagash River basin) to Chesuncook Lake (West Branch) connection which then enabled pulpwood to be sent by water route to the Millinocket mill.

The Eagle Lake and West Branch Railroad (E.L. & W.B. RR) line was engineered by, and the right-of-way was cleared by, Lacroix's Madawaska Company in 1925-26. In the spring of 1926, company crews started grading and laying the steel rails. Its Eagle Lake terminus was at the same site where, from 1901-1907, the tramway had pulled logs from the lake. The thirteen-mile railroad transported pulpwood to Umbazooksus Lake, where the *West Branch No. 2* then towed it across Chesuncook Lake and sent it down the West Branch to the Great Northern Paper Co. mills in Millinocket.

It was planned that GNP Co. would construct a five-mile long Chesuncook and Chamberlain Railroad (C. & C. RR) along the northern shore of Umbazooksus Lake which connected to Chesuncook Lake on the West Branch, but it was never built as a railroad. A roadway was constructed for hauling supplies, however. [55]

Locomotive Number 1 was built in the Schenectady Locomotive Works in New York, and Locomotive Number 2 was built in 1901 at Brooks Locomotive Works in Dunkirk, New York, and purchased by GNP Co. in 1928. [56]

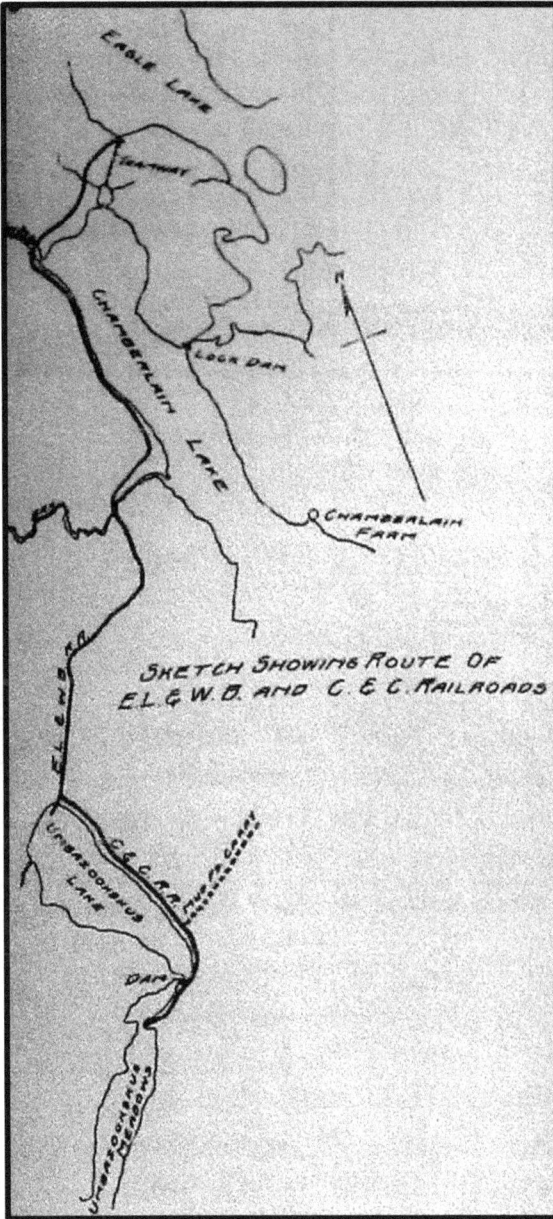

Sketch showing the E.L. & W.B. RR route, and the proposed Chamberlain and Chesuncook RR route

The Northern, November 1927, p. 4, Great Northern Paper Company Records, SpC MS 0210 Box 21, Raymond H. Fogler Library Special Collections Department, University of Maine, Orono, Maine

Lacroix purchased two 100-ton steam locomotives and 60 train cars to haul the logs on the railroad. He never actually operated the railroad, however, because GNP Co. purchased his operation in 1927. The E.L. & W.B. RR was very successful and operated through 1933. [57]

E.L. & W.B. RR Locomotive No. 2 (partial) transported from Canada on a snow sled
Terence F. Harper Collection

horse on the headworks winds in the slack of the boom for the
men who are feeding the conveyors

**At the Eagle Lake end of the railroad, a horse turns a capstan to draw the
boomed logs toward the railroad car conveyor.**
Great Northern Paper Company Records, SpC MS 0210, Raymond H. Fogler Library
Special Collections Department, University of Maine, Orono, Maine

A crew of men at Eagle Lake feed the logs into the conveyors which load the freight cars

The Northern, November 1927, p. 4, Great Northern Paper Company Records, SpC MS 0210, Box 21, Raymond H. Fogler Library Special Collections Department, University of Maine, Orono, Maine

Harkness had responsibility for conveyors in the GNP Co. pulpwood transport systems, and conveyors were extremely important to the efficiency of the railroad. As an example, loading the pulpwood on the E.L. & W.B. RR rails cars required two conveyors 225' long and 25' in height, each powered by a 40 hp diesel engine, to lift the pulpwood and dump it into the rail cars. Each of the three trains was made up of 12 cars each, and each car held 12 ½ cords. Using the conveyor system, each car could be loaded in 18 minutes. [58]

Note: O.A. Harkness's daughter, Betty Peters, recalled, in a telephone conversation with the author in November 2017, that the conveyor operators didn't feel the conveyors were safe to use, especially because the cars were moving slowly while the conveyors filled them. O.A. then sat beneath the operating conveyors to prove that they were designed and constructed for safe operation.

Two conveyors loading pulpwood logs into railcars.
Chesuncook Village Association website photo

Closeup of one of the conveyors used to load pulp onto railroad cars at Eagle Lake
www.maine.gov/dacf/parks/discover_history_explore_nature/history/Allagash/tramshtml

The railroad terminus at Eagle Lake showing that a third conveyor had been added.
Chesuncook Village Association website photo

Photo by M. Straus
The ninety ton oil burning steam engine makes a round trip from Eagle Lake
to Umbazookskus Lake every three hours

E.L. & W.B. RR in operation
Great Northern Paper Company Records, SpC MS 0210, Raymond H. Fogler
Library Special Collections Department, University of Maine, Orono, Maine

Lombard log hauler converted to run on rails
Terence F. Harper Collection

In his continual quest for improvements in motorized systems and equipment, O.A. Harkness oversaw the conversion by his crew of this 10-ton Lombard gasoline powered log hauler with wheels to fit on railroad tracks, with the modification work being done at the GNP Co. Greenville shop. This "tractor" was used to haul rails and other material needed for the E.L. & W.B. RR construction project. [59]

Abandoned Eagle Lake and West Branch Railroad locomotives
Photo by the Author 2017

When the Railroad stopped operating in 1933, after six years of operation, both locomotives were considered obsolete and were abandoned at their Eagle Lake terminus deep in the north Maine woods where they remain to this day. They were righted and restored through the efforts of the State of Maine Department of Agriculture, Conservation, and Forestry and the Allagash Alliance volunteers.

The remnants of the tramway (1901-1907), and the Eagle Lake and West Branch Railroad (1927-1933) terminus which re-used the tramway site on Eagle Lake, were listed on the National Register of Historic Places as the Tramway Historic District in 1979. They are part of the Allagash Wilderness Waterway, a Maine state park.

Although the development of railroads for log and pulpwood transportation took place in western Maine and other sections of the United States, the topography of the Maine north woods and the flow levels and directions of its river networks helped make river log driving feasible into the early 1970s.

Tanks at Greenville Junction.

The gasoline, kerosene and fuel oil storage tanks that have just been set by the G. N. P. Co. at Greenville Jct., have a capacity as follows: The three tanks for gasoline hold 34,000 gallons. The gas will all be handled through a nelectric registering pump. One tank for kerosene, 10,500 gallons; one tank for fuel oil, 18,000 gallons. The gasoline will be distributed from these tanks to Grant Farm tank, tank at Chesuncook Dam and all places on this side of the lake where gas is used. The White Tank Truck with a capacity of 1200 gallons is used for distributing. The crude oil will be used by the crude oil boats on Chesuncook Lake. There are two of these boats in use on Chesuncook Lake this season.

O. A. HARKNESS.

Fuel storage and distribution systems
The Northern, Great Northern Paper Company Records, SpC MS 0210, Box 21,
Raymond H. Fogler Library Special Collections Department, University of Maine,
Orono, Maine

It should be noted, too, that most GNP Co. motorized equipment required either gasoline, kerosene, or diesel (crude oil) fuel to keep it running. *The Northern* article (above) by O.A. Harkness describes GNP Co.'s tank farm's capacity of 62,500 gallons of fuel, and how it was allocated by fuel type. Distribution of fuel to the GNP Co. boats and sites of logging operations was accomplished using fuel delivery trucks.

As an example of the complexities of delivering fuel deep into the north woods, diesel fuel for the locomotives, for the engines powering the railroad car-loading conveyors, and for the power plants generating electricity at the E.L. & W. B. RR site was brought from the Greenville

Junction fuel depot to Chesuncook Lake dam by a Madawaska Company fleet of trucks. At the dam, the barrels were loaded onto a barge and taken to Umbazooksus Lake where they were unloaded and hauled by a small Plymouth yard engine leased from GNP Co. to the Railroad terminus. [60]

Chapter 11

The End of River Log Driving

The liabilities of the reliable transport of logs by water, such as drought, the necessity of springtime-only log runs due to the volume of water needed to move them along the river courses, and the increasing distance of logging operations from the lake/river log driving networks, made the option of transporting logs via train, and then by truck, ever more feasible as the future unfolded.

In 1976, the centuries-old Maine practice of using rivers to transport logs to saw and paper mills came to an end, the result of Mainers and the rest of America's changing views on water pollution, recreation, paper making and highway transportation. The change that had been coming for years finally arrived, and in the mid-1970s the last log drives came down the Kennebec River from Moosehead Lake and down the West Branch of the Penobscot River basin.

The federal Clean Water Act of 1972, sponsored by Maine's U.S. Senator Edmund Muskie, was ushered into law by the U.S. Congress. Muskie, who was from the paper mill town of Rumford, knew what the logging and paper mills added to the Maine economy but deplored what it did to the rivers. A Maine law that banned log driving by October 31, 1976, was approved by the legislature. It was the environmental impact of the industrial process water discharges from the paper mills along the river that brought the Clean Water Act into effect. The log drives contributed to environmental impacts, too, as bark from the logs that settled on the river bottoms caused pollution and removed habitat for fish spawning and drastically reduced populations of both fresh and salt water species. Logs in the rivers disrupted boaters, fishermen and canoeists. Paper manufacturers made the adjustment to transport logs by truck, building hundreds of miles of logging roads and constructing some new mills.

Pulpwood moving to a paper mill
Chesuncook Village Association website photo

**Log booms of pulpwood behind the North Twin Lake dam
on the "lower lakes" near Millinocket**
Great Northern Paper Company Records, SpC MS 0210, Raymond H. Fogler Library
Special Collections Department, University of Maine, Orono, Maine

Chapter 12

O.A. Harkness's U.S. Patents

Ever seeking to improve equipment and machinery, O.A. Harkness holds two patents for his inventions. The U.S. Patent Office files show that he was granted these patents while he was employed by GNP Co.

Patent for grouser

On June 15, 1926, he was granted patent #1,588,549 for the design of a grouser. Grousers are devices intended to increase the traction of continuous tracks, especially in loose material such as soil or snow. This is done by increasing contact with the ground with protrusions, like conventional tire treads or an athlete's cleated shoes. A likely application of this patent was to improve winter traction on the Lombard haulers.

Patent for a steering mechanism for twin engine tractors

O.A. Harkness's U.S. Patent #1,805,141 approved on May 31, 1931, was for a steering mechanism for a two-engine crawler tractor. The Lombard "New Twin" crawler tractor had an engine on each side, with brake and rear-end gears mounted independently.

Mr. Harkness insisted on the Lombard crawler tractor having exceptionally strong and rugged gears, cases and frame. He also wanted the machine to have proper balance, adequate speed, and the ability to travel through rough country and deep snow.

Harkness's invention of a steering system to simplify steering of this two-engine tractor resulted in remarkable flexibility in steering. A steering wheel like that of an automobile controlled the throttles of both engines. When the steering wheel was turned in either direction, it automatically throttled down the engine on that side, moving the tractor to the intended direction.

Harkness's Grouser Patent

Grousers are devices attached to continuous track vehicles like the Lombard A-O-L Twin Tractor which are intended to increase the traction, especially in loose material such as soil or snow, and on ice, by increasing ground contact with gripping protrusions.

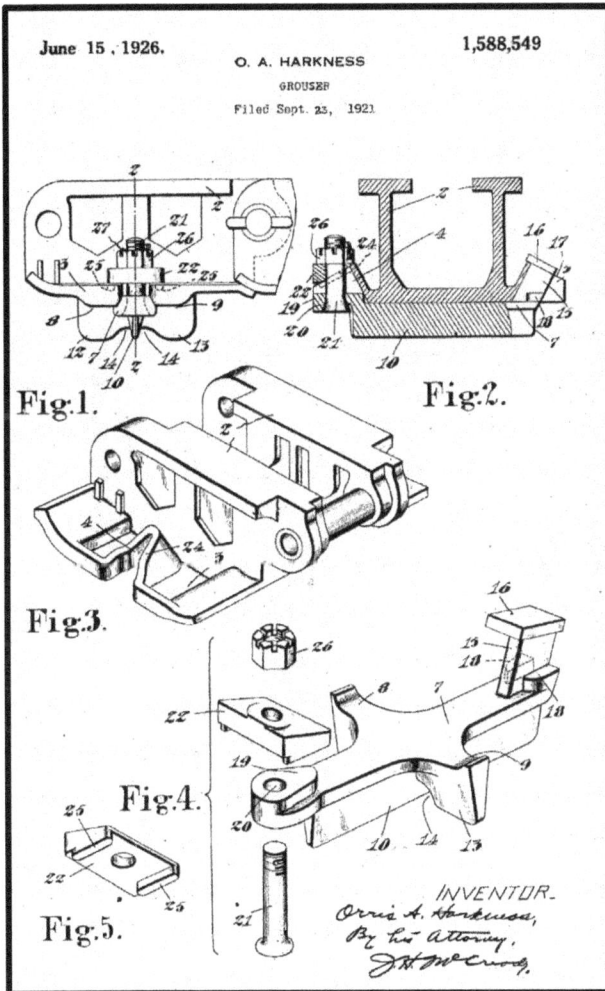

June 15 , 1926.

O. A. HARKNESS

GROUSER

Filed Sept. 23, 1921

1,588,549

Fig.1.

Fig.2.

Fig.3.

Fig.4.

Fig.5.

INVENTOR.

Orris A. Harkness,

By his attorney,

113

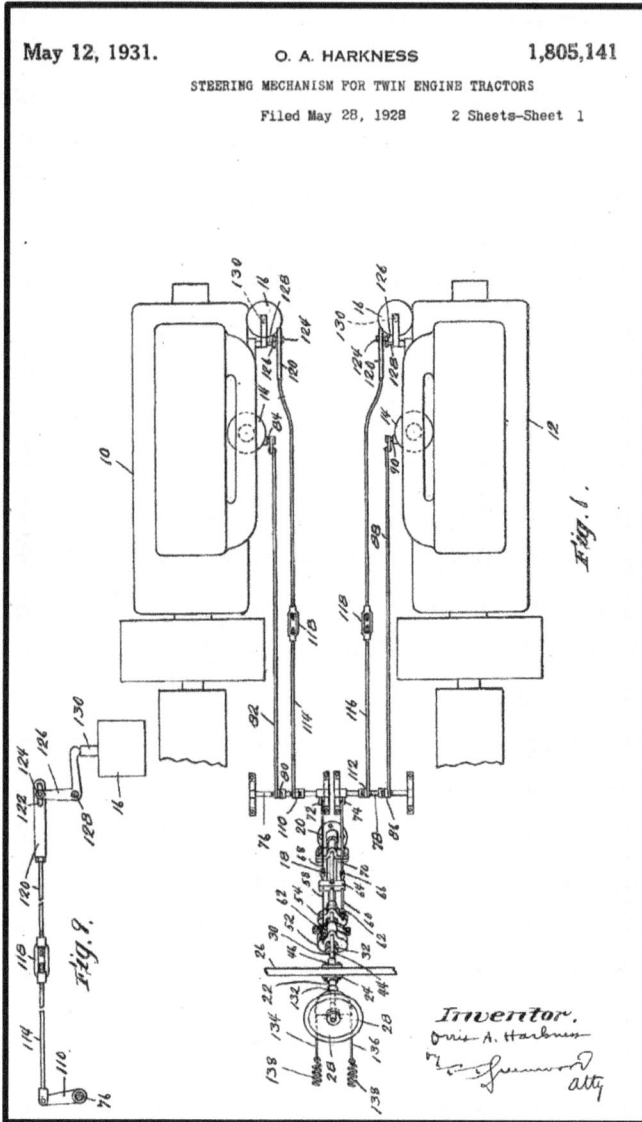

May 12, 1931.

O. A. HARKNESS

1,805,141

STEERING MECHANISM FOR TWIN ENGINE TRACTORS

Filed May 28, 1928

2 Sheets-Sheet 1

Harkness's Steering Mechanism Design for the A-O-L Twin Engine Tractors, by which crawler tractors could be steered with a steering wheel
U.S. Patent Office

Chapter 13

O.A. Harkness's Electric-Powered
Bucksaw Design Plan

O.A. Harkness designed an electric-powered bucksaw in 1932.

Early "Dragsaws," Powered Bucksaws, Bow Saws, and Chainsaws

In the 1920s there was great interest in devising mechanical saws to ease the work of timber and pulpwood harvesting, and in 1932 O.A. Harkness designed this electric-powered bucksaw.

Dragsaws were reciprocating power saws using a six-foot steel crosscut saw to buck logs to length and were the forerunners of the modern chainsaw. These mechanical bucksaws came into use in northern California and Oregon starting about 1909, where they were the first powered saws to be used in U.S. logging operations. The earliest models were powered by steam engines, but later models were gasoline powered. In theory they were portable, but some weighed 300 pounds. Prior to the popularization of the chainsaw during World War II, the dragsaw was a popular means of taking the hard work out of cutting wood. They were very reliable, rugged, and significantly more efficient than cutting (bucking) by hand. Some manufacturers included Multnomah, Vaughn, and Wee McGregor. (From Wikipedia)

Dragsaw example
Internet website photo

116

In terms of mechanics, it had long been known that continuous rotation in one direction was more efficient for cutting than the reciprocating back-and-forth motion of a handsaw or drag saw. Because a reciprocating saw must stop and reverse direction, it can never build up the cutter velocity that a rotating blade can. A circular blade isn't practical for tree-felling, though, because it can only cut half as deep as the saw blade is broad. Visionaries kept coming back to the concept of a linear blade with a line of teeth around its outer edges, something that could only be achieved with a loped chain running in a track. Experimental chain saws began to appear in both Sweden and British Columbia in 1919. [62]

Charles Wolf established the first viable chain saw firm in North America in 1920, manufacturing several models with a cutting chain which gripped a sprocket that propelled it on a custom fabricated bar. Such saws were usually electric-powered by a generator or electrified sawmill. The technology of gasoline-powered saws for logging operations developed at an astounding pace in the decade of 1925-35, though, and by the mid-1930s German companies Stihl and Dolmar were both producing high quality gas-powered chainsaws sold in Europe and in North America. [63]

GNP Co. file photographs show loggers using gas-powered saws in harvesting operations beginning about 1947.

O.A. Harkness's Electric Powered Bucksaw Design Plan

Showing Davey Air Saw.

No. 11,113   Feb. 11, 1944

Property of Great Northern Paper, Inc.

GNP Co.'s Davey compressed air saw, 1944
Chesuncook Village Association website photo

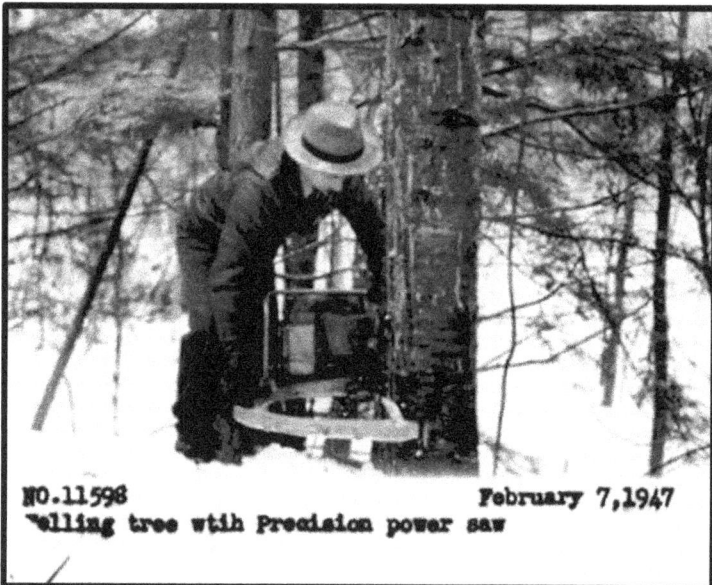

NO.11598   February 7,1947
elling tree wtih Precision power saw

GNP Co. 1947 Precision "bow" chainsaw, manufactured in Montreal, Quebec
Chesuncook Village Association website photo

Property of Great Northern Paper, Inc.

**GNP Co. photo of Mall Tool Company bow saw (ca. 1951)
made in Milwaukee, Wisconsin**
North Maine Woods, Inc. website photo

With a recoil starter instead of a rope that had to be manually wound each time, the Mall gas-powered saws were popular in the lumber industry. [64]

GNP Co.'s experimental "Sally Saw"
Chesuncook Village Association website photo

Produced by Cummings Machine Works, Boston, Massachusetts, in the 1940s, the Sally Saw used a gas-powered motor to operate a circular saw at the end of a drive shaft. They weighed about 75 lbs.

Harkness's 1932 design for an electric-powered bucksaw solved weight issues, but didn't incorporate chain saw efficiency or meet the need for a gasoline-powered, easily portable saw. There's no evidence that Harkness's electric bucksaw was ever fabricated.

The photographic record shows, however, that GNP Co. was regularly experimenting in the 1940s with chainsaws using newly-developed technologies.

Epilogue

Philip T. Coolidge, in his 812-page masterwork *History of the North Maine Woods* says this about O.A. Harkness: "I must pay respects to the late O.A. Harkness, at first employed by the Eastern Manufacturing Company, later by the Great Northern. In the pages of *The Great Northern* [which was published from 1921-1928] his contributions give an invaluable record of events in the lumbering and pulpwood operations." [65]

Certainly Harkness's combination of technical prowess and the ability to convey it in literary form was a rare and valued ability.

O.A. Harkness, as developer of the boom jumper work boat, was able to fulfill the need for a boat that could operate in water filled with four-foot long pulpwood held together by floating log booms which needed to be crossed. The boom jumpers were used primarily to do errands and light work around booming-out places, and bringing up and placing slack booms. The fact that they remained an essential part of GNP Co.'s operational fleet from 1914 to the end of log drives in 1972, a remarkable span of 58 years, speaks to the longevity of the concept and its successful, evolving execution.

Gil Gilbert reported in his *Allagash: A Journey Through Time on Maine's Legendary Wilderness Waterway* that in 1925 a decision-making group was formed to make a canoe trip down the East Branch of the Penobscot to take one last look at that Branch of the river before committing to the West Branch for moving logs to the Millinocket mills. With the combination of their long experience with and understanding of operating costs, marketing, pulpwood transportation systems, and the geography of the north woods, GNP Co.'s Vice President Fred A. Gilbert, President of the Madawaska Company Edouard "King" Lacroix, and GNP Co. Superintendent of Motor Power O.A. Harkness, conceived an even more visionary and ambitious scheme: they planned a logging railroad through the wilderness from Eagle Lake to Umbazooksus on the West Branch. Harkness clearly added significantly to the work of this high-level task force, which concluded that preparing the East Branch for log drives was cost-prohibitive due to several large drops in the river and the fact that several large dams would have to be built at strategic points. Their findings sealed the future of the West Branch as the primary route for moving pulpwood to the GNP Co. paper mills, even over the option of the Bangor & Aroostook Railroad line which stretched well into northern Maine.

A 1927 article in *The Northern* on the subject of "Admiral Harkness' Inland Fleet" begins thusly: "Of all the men in the Spruce Woods Department, none travels farther or is in greater demand than O.A. Harkness. He may be seen anywhere but not for long as he is constantly covering the whole territory where gasoline oil or coal is used for motive power. It is difficult to hold him long enough for and [*sic*] interview...."

It seems incredible, too, that Harkness could drive 20,000 to 25,000 miles in a five-month summer/fall season on what would be considered today as a very primitive unpaved road network. Yet as *The Northern* article of May 1925 marveled, "It is no wonder that his Franklin [auto] registers more than 25,000 miles or more during a season. He is in great demand, and called in every direction." Similarly, the Stutz Motor Car article reprinted in Appendix I reports, "From conversation with Mr. Harkness the facts developed that the Great Northern Paper Co. had purchased the Stutz Victoria coupe on May 2, 1926, and in the short period of less than five months' time had been driven over 20,000 miles over all manner of roads and highways." These observations document his high level of personal energy and his commitment to accomplishing the work for which he was ideally suited.

Additionally, the GNP Co. employee newsletter article written on the occasion of the June 6, 1964, christening of the steel tug *O.A. Harkness* gives him this high praise: "Though he lacked much in formal education, he was a mechanical genius of the highest order. ... A list of his inventions and improvements on equipment would take pages to list. Above all, he was a man who inspired others and who earned the respect of all who came into contact with him."

Similarly, the poem written by Harkness's fellow GNP Co. employees for his 80th birthday (see Appendix I) reflects the esteem and respect in which he was held by his former co-workers for both his mechanical systems prowess, and as well as noting his penchant for smoking cigars and having a "modest" personality. (See Appendix I)

Orris Albert Harkness, then, was a highly energetic and widely recognized mechanical genius. He was also trusted and respected by the

GNP Co. managers for the insights his mechanical prowess provided to logging transport systems, and by the logging work force for conveying practical operational information in ways that could inspire them. Throughout his long career, he provided ongoing and valued perspective, understanding, and insights to GNP Co.'s owners and management leaders for making optimal decisions for the boat fleet's vessels and functions, about which he was especially proud, and a huge array of other mechanical equipment and operations.

The title of inland fleet "Admiral" characterizes him perfectly, as does the appellation of mechanical "Wizzard."

Appendix I

Harkness Family Collection

Vinton Orris "V.O." Harkness, Jr., grandson of O.A. Harkness
2017 Family Photo

V.O. Harkness graciously provided the following Harkness family items of memorabilia for this Appendix.

Sketch of steam launch *RAY.* 1897
V.O. Harkness, Jr. Collection.
Photo by the author 2017

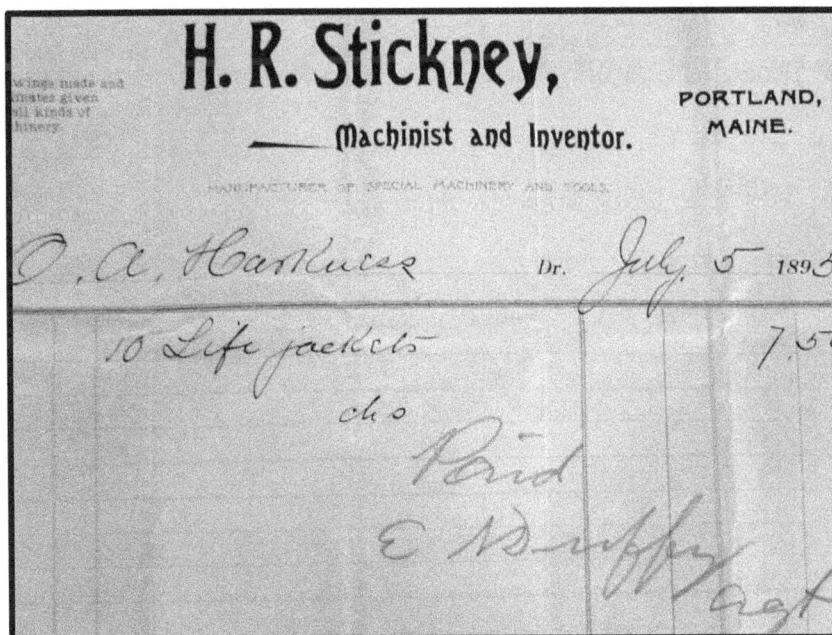

1895 receipt for 10 life jackets for the *RAY.*
V.O. Harkness, Jr. Collection.
Photo by the author 2017

These four logging tow boat models were built about 2002 by "Ship Shape" model builder Peter Templeton, 296 Pritham Avenue, Greenville Junction, Maine, on a commission from V.O. Harkness, Jr. The *George A. Dugan* and *West Branch No. 2* were designed by O.A. Harkness and built under his supervision.

Towboat *F.W. Ayer*, 1883
V.O. Harkness, Jr. Collection
Photo by the author 2017

Towboat *George A. Dugan*, 1902
V.O. Harkness, Jr. Collection
Photo by the author 2017

126

Towboat *West Branch No. 2*, 1927
V.O. Harkness, Jr. Collection
Photo by the author 2017

Towboat *O.A. Harkness*, 1964
V.O. Harkness, Jr. Collection
Photo by the author 2017

O.A. Harkness
Terence F. Harper Collection

**O.A. Harkness at Megunticook Farm on Fernald's Neck,
Lincolnville, Maine, ca. 1930**
V.O. Harkness, Jr. Collection

**O.A. Harkness, his son Vinton Orris Harkness, Sr.,
and grandson Vinton Orris "V.O" Harkness, Jr., 1942**
V.O. Harkness, Jr. Collection

O.A. Harkness in 1946
V.O. Harkness, Jr. Collection

O. O. Harkness,
Camden,
Maine,
1889

Given To V. D. HARKNESS, Jr., by his
grandfather, O. A. HARKNESS — Summer 1941-

2 vols. [illegible], m. e.

2520

15

MODERN STEAM PRACTICE,

ENGINEERING AND ELECTRICITY:

A GUIDE TO

APPROVED METHODS OF CONSTRUCTION AND
THE PRINCIPLES RELATING THERETO.

WITH

EXAMPLES, PRACTICAL RULES, AND FORMULÆ.

BY

JOHN G. WINTON, ENGINEER,

AUTHOR OF "MODERN WORKSHOP PRACTICE."

ASSISTED BY

W. J. MILLAR, C.E.,

Secretary of the Institution of Engineers and Shipbuilders in Scotland;
Author of "Principles of Mechanics," &c.

*ILLUSTRATED BY 900 ENGRAVINGS IN THE TEXT
AND A SERIES OF SEPARATE PLATES.*

PHILADELPHIA:

GEBBIE & CO.

1883.

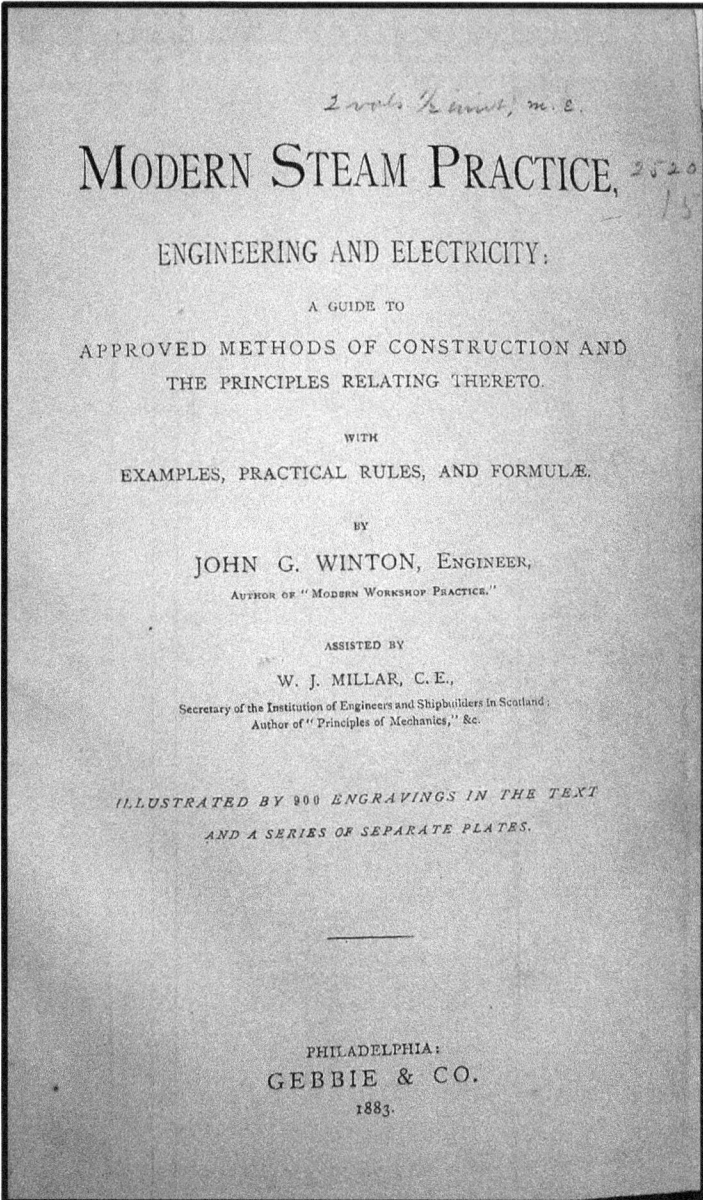

O.A. Harkness's steam engine "bible", a two-volume set which he gave to his grandson V.O. Harkness, Jr., in the summer of 1941
V.O. Harkness, Jr. Collection
Photo by the author 2017

BANGOR DAILY NEWS, SATURDAY, JANUARY, 29, 1927.

Unparalleled Motor Feat is Performed by Stutz Victoria

MORE ON

Orris A. Harkness, Chief Engineer of Mechanical Department of Great Northern Paper Co., Drove Stutz Victoria Coupe 20,000 Miles in Less Than Five Months Without Single Mechanical Adjustment—Local Man Pushed Car Over Seemingly Unpassable Roads, Day In Day Out, and Never Once Experienced the Slightest Difficulty With the Stutz's Mechanism

Orris A. Harkness Standing Beside Stutz Victoria Coupe, Which Performed Valiant Service During Summer Months.

Harkness logs 20,000 auto miles in 5 months
V.O. Harkness, Jr. Collection (See text excerpt on the following page)

133

Note: Although this January 29, 1927, newspaper article is primarily advertising for the Stutz Motor Car Company, it does contain some interesting information about O.A. Harkness's responsibilities at GNP Co. and his incredible work ethic.

"Quite recently Orris A. Harkness, chief of the mechanical department in the spruce wood division of the Great Northern Paper Co., drove the Stutz coupe into the J.M. Norris Motor Co. establishment, and while he stated that the car was in perfect running condition, he nevertheless explained that it had been subjected to extremely hard usage during the summer months.

From conversation with Mr. Harkness the facts developed that the Great Northern Paper Co. had purchased the Stutz Victoria coupe on May 2, 1926, and in the short period of less than five months' time had been driven over 20,000 miles over all manner of roads and highways, without necessitating one single mechanical adjustment.

His work as chief engineer of the mechanical department of the Great Northern Paper Company, an institution comprising resources of millions of dollars and considered one of the largest pulp operating concerns in the United States, required him to be in widely scattered towns and cities or wherever trouble occurred within the scope of the company's operations in all sections of the state. And to be in several sections of the state in one day required hard and fast driving and subjected the motor car to the hardest treatment imaginable."

O.A. Harkness's daughter Betty Harkness Peters told the author that O.A.'s first car was a REO. His favorite, though, was a Franklin with an air-cooled engine which could climb steep hills without any down-shifting. She remembers Fords, too, and the back windows were always filled with rolls of blueprints. He often carried a spare spring in his cars, and always checked his tires before leaving the home in Veazie. During hunting season, he carried a double-barreled shotgun and sometimes his rifle.

She recalled that sometimes when he left early in the morning his breakfast was a whole raw egg dropped into a glass of milk! [A protein concentrated high-energy way to start one's day.]

His office was at Greenville Junction ... turn left after crossing the railroad tracks. Every Saturday morning, he went to the GNP office in Bangor, which was located on the top floor of the Eastern Trust and Banking Company. Then he'd have his car serviced at the Webber Motor Co. on Hammond Street [owned by Bangor businessman Alburney E. "Allie" Webber]. His final Saturday morning stop was to pick up fresh fish at a market on Broad Street in downtown Bangor.

Birthday Greetings to Orris - July 6th 1949

It was a gala day in Lincolnville just 80 yrs ago today
when Mamma & Pappa Harkness had cause to kneel & pray
A little dark-haired youngster, with great big eyes of brown
Was born to a sweet mother in that little sea side town.
They called him little Orris, a name now liked so well
And his many, many friends think Orris is just swell
His little home on Penobscot Bay where ships he could often see
Convinced him in his early life that a ship builder he would be
Yachts & boats designed by Orris were registered far & many
And the sinkings of his crafts designed, he will tell you were
      not any.
At 44 and a swell looking guy, he came with the GNP
He smoked cigars, yes several a day, & he wore a
      little goatee.
For 36 years as Master Mechanic, Orris has shown his art
And in the progress of the GNP, he has taken an active part.
It's easy rhyming lines for Orris & we could go on
      rhyming more.
But Orris is a modest soul, & the poor man might
      get sore.
So we are wishing him more birthdays yes, many
      more to be
It's the nicest wish that we could wish say his
      friends from the GNP.

Roger - the above poem was written on the back of a photo taken of all the men
at my father's (ORRIS') 80th birthday celebration in Greenville, all of the
men signed it. Betty Harkness Peters

Poem to O.A. Harkness from GNP Co. Employees
at his 80th Birthday Celebration in Greenville
Betty Harkness Peters Collection

Appendix II

Great Northern Paper Company Boat Index

GNP Co. Boat No.	Designer	Builder	Year	Function	Length in Ft.
"When he [Harkness] became employed the GNP in 1915, there were seven motorized boats in use. Five were Atlantic dories with Atlantic engines with horsepower ranging from 8 to 15. No. 3 and No. 10 were the other boats." (*The Northern*, September 1928)					
GNP Co. No. 1					
GNP Co. No. 2					
GNP Co. No. 3		Camden Anchor-Rockland Machine Company	1912	Workboat	37
GNP Co. No. 4		(No. 4 through No. 9 were Atlantic dories with Atlantic		Atlantic Dory	20 est.
GNP Co. No. 5		engines which were built by the Lunenburg Foundry		Atlantic Dory	20 est.
GNP Co. No. 6		in Lunenburg, Nova Scotia)		Atlantic Dory	20 est.
GNP Co. No. 7				Atlantic Dory	20 est.
GNP Co. No. 8				Atlantic Dory	20 est.
GNP Co. No. 9				Atlantic Dory	20 est.
GNP Co. No. 10	Fred Sawyer	GNP's Greenville Machine Shop, Fred Sawyer of Greenville	1912	First Boom Jumper	30
GNP Co. No. 11	O.A. Harkness	Cobb Brothers, Brewer	1916	Boom Jumper	35
GNP Co. No. 12	O.A. Harkness			Boom Jumper	
GNP Co. No. 13	O.A. Harkness			Boom Jumper	
GNP Co. No. 14	O.A. Harkness			Boom Jumper	
GNP Co. No. 15	O.A. Harkness			Boom Jumper	
GNP Co. No. 16	O.A. Harkness			Boom Jumper	
GNP Co. No. 17	O.A. Harkness	Cobb Brothers, Brewer	1919	Boom Jumper	30
GNP Co. No. 18	O.A. Harkness	Cobb Brothers, Brewer	1919	Boom Jumper	30
GNP Co. No. 19	O.A. Harkness	William St. Germaine, Greenville	1920	Boom Jumper	30
GNP Co. No. 20	O.A. Harkness	Cobb Brothers, Brewer		Boom Jumper	30
GNP Co. No. 21	O.A. Harkness	GNP's Greenville Machine Shop/C.H. Ingalls, Master Builder	1921	Towboat	58
GNP Co. No. 22	O.A. Harkness	Cobb Brothers, Brewer	1923	Boom Jumper	30
GNP Co. No. 23	O.A. Harkness	Cobb Brothers, Brewer	1923	Boom Jumper	30
GNP Co. No. 24	O.A. Harkness	Cobb Brothers, Brewer	1923	Boom Jumper	30
GNP Co. No. 25	O.A. Harkness	Cobb Brothers, Brewer	1923	Boom Jumper	30
GNP Co. No. 26	O.A. Harkness				
GNP Co. No. 27	O.A. Harkness	GNP's Greenville Machine Shop	1924	Boom Jumper	30
GNP Co. No. 28	O.A. Harkness	GNP's Greenville Machine Shop	1924	Boom Jumper	30
GNP Co. No. 29	O.A. Harkness	GNP's Greenville Machine Shop	1924	Boom Jumper	30
GNP Co. No. 30	O.A. Harkness	GNP's Greenville Machine Shop	1925	Boom Jumper	30
GNP Co. No. 31	O.A. Harkness	GNP's Greenville Machine Shop	1927	Boom Jumper	32
GNP Co. No. 32	O.A. Harkness	GNP's Greenville Machine Shop	1927	Boom Jumper	32

GNP Co. Boat No.	Designer	Builder	Year	Function	Length in Ft.
GNP Co. No. 33	O.A. Harkness	GNP's Greenville Machine Shop			
GNP Co. No. 34	O.A. Harkness	GNP's Greenville Machine Shop			
GNP Co. No. 35	O.A. Harkness	GNP's Greenville Machine Shop			
GNP Co. No. 36	O.A. Harkness	GNP's Greenville Machine Shop			40 est.
GNP Co. No. 37	O.A. Harkness	GNP's Greenville Machine Shop	1928	Boom Jumper	32 est.
GNP Co. No. 38	O.A. Harkness	GNP's Greenville Machine Shop			
GNP Co. No. 39	O.A. Harkness	GNP's Greenville Machine Shop			
GNP Co. No. 40	O.A. Harkness	GNP's Greenville Machine Shop		Boom Jumper	
GNP Co. No. 41	O.A. Harkness	GNP's Greenville Machine Shop	ca. 1945	Boom Jumper	40 est.
GNP Co. No. 42	Unknown			Boom Jumper	40 est.
GNP Co. No. 43	Unknown				
GNP Co. No. 44	Unknown				
O.A Harkness retired in 1950, and GNP Co. then moved to steel hulled boom jumpers designed by Naval architect Geerd Hendel, Camden, Maine					
GNP Co. No. 45	Geerd Hendel	General Foods Corp. Shipyard Rockland, Maine	1951	Boom Jumper	30
GNP Co. No. 46	Geerd Hendel	General Foods Corp. Shipyard	1952	Boom Jumper	30
GNP Co. No. 47	Geerd Hendel?				
GNP Co. No. 48	Geerd Hendel?				
GNP Co. No. 49	Geerd Hendel	General Foods Corp. Shipyard	1954	Boom Jumper	30
GNP Co. No. ___	Geerd Hendel	Goudy & Stevens &Hodgdon Bros. Partnership, hull #132	1956-57	Boom Jumper/Tug	35
GNP Co. No. 57	Geerd Hendel	Goudy & Stevens &Hodgdon Bros. Partnership, hull #143	1956-57	Boom Jumper/Tug	35
GNP Co. No. ___	Geerd Hendel	Goudy & Stevens & Hodgdon Bros. Partnership, hull #146		Workboat	28
GNP Co. No.___	Geerd Hendel	Goudy & Stevens & Hodgdon Bros. Partnership, hull #194	1962	Workboat, With tunnel drive	30
GNP Co. No. ___	Geerd Hendel	Goudy & Stevens & Hodgdon Bros. Partnership, hull #197	1963	Workboat	35
West Branch No. 1	likely O.A. Harkness	Great Northern Paper Co.	early 1920s	Towboat	90 est.
West Branch No. 2	O.A. Harkness	Great Northern Paper Co.	1927	Towboat	90 est.
West Branch No. 3	O.A. Harkness	Great Northern Paper Co.	1948	Towboat	91
William Hilton	Geerd Hendel	Goudy & Stevens, East Boothbay, hull #192	1961	Towboat	70
O.A. Harkness	Geerd Hendel	Goudy & Stevens, East Boothbay, hull #199	1964	Towboat	70

Reference Sources

1. John E. McLeod, *The Northern – The Way I Remember*, Condensed from a history by John E. McLeod, Millinocket, Maine: Great Northern Paper Company, 1982, 38 and 85.
2. Paul K. McCann, *Timber! The Fall of Maine's Paper Giant ... A chronicle of Great Northern Paper Company in the 1970s and 1980s*, ©Great Northern Paper Company, 1994, 2.
3. *The Northern*, Magazine of the Great Northern Paper Company, May 1925, 8. Great Northern Paper Company Records, SpC MS 0210 Box 21, Raymond H. Fogler Library Special Collections Department, University of Maine, Orono, Maine
4. James O. McKinsey, *Organization of the Walworth Manufacturing Company*, University of Chicago Press, in the Journal of Political Economy, Vol. 30, No. 3, 1922, 420-421.
5. Wayne E. Reilly, "Fred Ayer and Eastern Manufacturing" from *Remembering Bangor, The Queen City Before the Great Fire*, The History Press, Charleston, SC, 2009.
6. John E. McLeod, *The Northern – The Way I Remember*, 29 and 31.
7. *The Northern*, May 1928, 4.
8. *The Northern*, December 1926.
9. David C. Smith, *A History of Lumbering in Maine 1861-1960*, University of Maine Studies No. 93, University of Maine Press, Orono, Maine 1972, 83.
10. Charles Glaster, *The West Brancher*, Vantage Press, Inc. New York, New York, 1970, 113.
11. *The Northern*, February 1927, 7.
12. *The Northern*, November 1927, 6.
13. *The Northern*, May 1921, 3.
14. *Maine Lakes Steamboat Album*, Downeast Enterprises, Inc., Camden, Maine, 1976, 11-12.
15. *The Northern*, December 1926, 3-4.
16. *The Northern*, May 1928, 3.
17. Laura Gagnon, *Chesuncook Memories*, Chesuncook Village Association, 1989, 78-79.

18. Moosehead Historical Society newspaper article, Accession #81.4.41
19. *The Northern*, September 1928, 3.
20. Cecil Max Hilton, "Equipment and Methods Used at Pulpwood Operations Upon the West Branch of the Penobscot River in Maine, 1935-40," University of Maine Thesis, 1940
21. *The Northern*, September 1928, 4.
22. Lew Dietz, "Special Breed of Boats Worked Maine's Lakes," *National Fisherman*, May 1967, 18A.
23. Bertram B. Snow, *The Main Beam*, published by the Rockland Historical Society, Rockland, Maine, 2005, 456.
24. Ibid, 452.
25. *The Northern,* February 1927, 6.
26. Glaster, 139.
27. *The Northern*, September 1928, 4-5.
28. John Gardner, *The Dory Book,* International Marine Publishing Company, Camden, Maine, 1978, 135.
29. *Bangor Daily News*, "The O.A. Harkness," September 5, 1972.
30. "The Loss of the Tug O.A. Harkness," *The Maine Citizen*, January 16, 1992.
31. Maine Department of Agriculture, Conservation and Forestry, DACF Home → Bureaus & Programs → Bureau of Parks and Lands → Discover History & Explore Nature → History & Historic Sites → Telos Dam and Cut
32. Smith, 396.
33. *The Northern*, November 1927, 6.
34. Gil Gilpatrick, *Allagash: A Journey Through Time on Maine's Legendary Wilderness Waterway*, DeLorme Publishing Co., Freeport, Maine 1983, 120.
35. Smith, 396.
36. *The Northern*, November 1927, 14.
37. Gilpatrick, p. 119.
38. *Maine Lakes Steamboat Album*, 11.
39. Gilpatrick, 119.

40. Lore A. Rogers and Caleb W. Scribner, *The Log Haulers*, Patten Lumbermen's Museum, Patten, Maine, The American Society of Mechanical Engineers, 1982, 2-5.

41. Ibid, 3.

42. Terence F. Harper, *Tracks Through time, A History of the Maine Forest & Logging Museum's Steam Lombard*, May 2015, 3.

43. *The Northern*, December 1927, 5.

44. *The Northern*, December 1927, 15.

45. Terence F. Harper, *A Most Remarkable Machine, A History of the Lombard Log Hauler*, 2009, 10-11.

46. Rogers and Scribner, 3.

47. *The Northern*, June 1922, 13.

48. Richard Fraser and Nancy Fraser, *A History of Maine Built Automobiles & Motorized Vehicles 1834-1934*, Richard and Nancy Fraser, PO Box 39, E. Poland, ME 04230

49. *The Northern*, April 1927, 7 and 15.

50. Ibid.

51. Terence F. Harper, "Old Twin", "New Twin" The Lombard A-O-L, www.maineforestandloggingmuseum.org, February 2017

52. Gilpatrick, 142.

53. Richard William Judd and Patricia A. Judd, *Aroostook: A Century of Logging in Northern Maine*, University of Maine Press, 1989, 184.

54. Gilpatrick, 142-143.

55. Ibid.

56. Maine Department of Agriculture, Conservation and Forestry, DACF Home → Bureaus & Programs → Bureau of Parks and Lands → Discover History & Explore Nature → History & Historic Sites → Allagash History, 5.

57. Maine Department of Agriculture, Conservation and Forestry, *The Eagle Lake and West Branch Railroad History: Discover History and Explore Nature: State Parks and Public Lands*, 1.

58. Ibid, 2.

59. Terry Harper, *Railway Preservation News*, online post by Terry Harper dated May 8, 2007.

60. DACF Home → Bureaus & Programs → Bureau of Parks and Lands → Discover History & Explore Nature → History & Historic Sites → The Eagle Lake & West Branch Railroad, 1

61. Doug Harlow, *"The Last Log Drive: When a Maine Way of Life Came to an End," Portland Press Herald,* February 20, 2016.

62. David Lee, *CHAINSAWS, A History,* ©Harbor Publishing Co. Ltd., Madeira Park, British Columbia, Canada, 2006, 22-24.

63. Ibid, 37.

64. Ibid, 111.

65. Philip T. Coolidge, *History of the Maine Woods*, Furbush-Roberts Printing Company, Bangor, Maine, 1963, 618.

ABOUT THE AUTHOR

Roger Moody graduated from Camden High School in 1961, from the University of Maine with a Bachelor's Degree in History and Government in 1965, and then served two years of active duty as a Lieutenant in the U.S. Army Quartermaster Corps. In 1969, he received a Master of Public Administration degree from The Maxwell School of Citizenship and Public Affairs at Syracuse University, where his concentration was in the Metropolitan Studies Program.

His long-term public service career has included the administrative assignments in State of Connecticut's Department of Community Affairs, and in the Town of East Hartford, Connecticut. He has served as municipal manager for the City of Ellsworth, Maine, and for the Town of Camden, Maine, and as School Department Business Manager for the City of Bangor, Maine. He was elected to two terms as Knox County Commissioner, 2008-12 and 2013-16.

www.ingramcontent.com/pod-product-compliance
Lightning Source LLC
Chambersburg PA
CBHW021931190326
41519CB00009B/987